ERNEST HEMINGWAY

THE KILLERS

AND OTHER SHORT STORIES

Edited by

Dr. Georg Schnöckelborg

Oberstudiendirektor

FERDINAND SCHÖNINGH · PADERBORN

München · Wien · Zürich
Printed in Germany
Herstellung: Ferdinand Schöningh, Paderborn

ISBN 3-506-43116-1

CONTENTS

ABOUT THE AUTHOR

Ernest Miller Hemingway was born at Oak Park near Chicago, on July 21st, 1898, the son of a country doctor. Young Ernest spent most of his vacations in the Michigan woods. After he had graduated from high school in 1917, he got a job as a reporter in Kansas City, where he got to know prize-fighters and gunmen. His war experiences took him to France and Italy (World War I), where he served as an ambulance driver and was severely wounded, to Spain (Civil War, 1937) and again to France (1944). In 1954, he was awarded the Nobel Prize for Literature "for his powerful, style-forming mastery of the art of modern narration." Also it was felt in certain quarters of the Swedish Academy that the well-deserved prize should be awarded before "the man really kills himself." During his last hunting trip to Africa in January 1954, he went through two air crashes in the African jungle and nearly perished. He died on July 2nd, 1961, very likely killed himself, and lies buried at the base of Idaho's Sawtooth mountains.

His best-known books include *Fiesta* (called in the U.S. *The Sun also Rises,* 1926), *A Farewell to Arms* (1929), *To Have and Have Not* (1937), *For Whom the Bell Tolls* (1940), *The Snows of Kilimanjaro* (1948), *Across the River and into the Trees* (1950), and the novelette *The Old Man and the Sea* (1952).

From the two collections of short stories, *In Our Time* and *Men without Women,* a selection is presented in this little volume. They contain the magic of his pen as much as his longer works and reveal his favourite emotions: utter despair, hopelessness, futility, love, pride of battle, the courage not to bow before hostile circumstances.

About his style Maurice Baudin jr. writes in 'Contemporary Short Stories' ('The American Heritage Series'): "Hemingway is alone among major writers in the extent to which he has combined the active (boxing, fishing, hunting and war) and the writing lives. Disliking rhetorical writing and preferring the concrete symbol to the vague abstraction, and influenced by Stein, Turgenev, Anderson and Twain, among others, he has evolved the style that now bears his name: the style that is simple, direct, hard-hitting and precise, that employs no word its writer would not use in conversation, that deals in understatement and meaningful omission."

G. Sch.

THE KILLERS

The door of Henry's lunch-room opened and two men came in. They sat down at the counter.

'What's yours?' George asked them.

'I don't know,' one of the men said. 'What do you want to eat, Al?'

'I don't know,' said Al. 'I don't know what I want to eat.'

Outside it was getting dark. The street light came on outside the window. The two men at the counter read the menu. From the other end of the counter Nick Adams watched them. He had been talking to George when they came in.

'I'll have a roast pork tenderloin with apple sauce and mashed potatoes,' the first man said.

'It isn't ready yet.'

'What the hell do you put it on the card for?'

'That's the dinner,' George explained. 'You can get that at six o'clock.'

George looked at the clock on the wall behind the counter.

'It's five o'clock.'

'The clock says twenty minutes past five,' the second man said.

'It's twenty minutes fast.'

'Oh, to hell with the clock,' the first man said. 'What have you got to eat?'

'I can give you any kind of sandwiches,' George said. 'You can have ham and eggs, bacon and eggs, liver and bacon, or a steak.'

'Give me chicken croquettes with green peas and cream sauce and mashed potatoes.'

'That's the dinner.'

'Everything we want's the dinner, eh? That's the way you work it.'

'I can give you ham and eggs, bacon and eggs, liver —'

'I'll take ham and eggs,' the man called Al said. He wore a derby hat and a black overcoat buttoned across the chest. His face was small and white and he had tight lips. He wore a silk muffler and gloves.

'Give me bacon and eggs,' said the other man. He was about the same size as Al. Their faces were different, but they were dressed like twins. Both wore overcoats too tight for them. They sat leaning forward, their elbows on the counter.

'Got anything to drink?' Al asked.

'Silver beer, bevo, ginger-ale,' George said.

'I mean you got anything to *drink?*'

'Just those I said.'

'This is a hot town,' said the other. 'What do they call it?'

'Summit.'

'Ever hear of it?' Al asked his friend.

'No,' said the friend.

'What do you do here nights?' Al asked.

'They eat the dinner,' his friend said. 'They all come here and eat the big dinner.'

'That's right,' George said.

'So you think that's right?' Al asked George.

'Sure.'

'You're a pretty bright boy, aren't you?'

'Sure,' said George.

'Well, you're not,' said the other little man. 'Is he, Al?'

'He's dumb,' said Al. He turned to Nick. 'What's your name?'

'Adams.'

'Another bright boy,' Al said. 'Ain't he a bright boy, Max?'

'The town's full of bright boys,' Max said.

George put the two platters, one of ham and eggs, the other of bacon and eggs, on the counter. He set down two side-dishes of fried potatoes and closed the wicket into the kitchen.

'Which is yours?' he asked Al.

'Don't you remember?'

'Ham and eggs.'

'Just a bright boy,' Max said. He leaned forward and took the ham and eggs. Both men ate with their gloves on. George watched them eat.

'What are *you* looking at?' Max looked at George.

'Nothing.'

'The hell you were. You were looking at me.'

'Maybe the boy meant it for a joke, Max,' Al said.

George laughed.

'*You* don't have to laugh,' Max said to him. '*You* don't have to laugh at all, see?'

'All right,' said George.

'So he thinks it's all right.' Max turned to Al. 'He thinks it's all right. That's a good one.'

'Oh, he's a thinker,' Al said. They went on eating.

'What's the bright boy's name down the counter?' Al asked Max.

'Hey, bright boy,' Max said to Nick. 'You go around on the other side of the counter with your boy friend.'

'What's the idea?' Nick asked.

'There isn't any idea.'

'You better go around, bright boy,' Al said. Nick went around behind the counter.

'What's the idea?' George asked.

'None of your damn business,' Al said. 'Who's out in the kitchen?'

'The nigger.'

'What do you mean the nigger?'

'The nigger that cooks.'

'Tell him to come in.'

'What's the idea?'

'Tell him to come in.'

'Where do you think you are?'

'We know damn well where we are,' the man called Max said. 'Do we look silly?'

'You talk silly,' Al said to him. 'What the hell do you argue with this kid for? Listen,' he said to George, 'tell the nigger to come out here.'

'What are you going to do to him?'

'Nothing. Use your head, bright boy. What would we do to a nigger?'

George opened the slit that opened back into the kitchen. 'Sam,' he called. 'Come in here a minute.'

The door to the kitchen opened and the nigger came in. 'What was it?' he asked. The two men at the counter took a look at him.

'All right, nigger. You stand right there,' Al said.

Sam, the nigger, standing in his apron, looked at the two men sitting at the counter. 'Yes, sir,' he said. Al got down from his stool.

'I'm going back to the kitchen with the nigger and bright boy,' he said. 'Go on back to the kitchen, nigger. You go with him, bright boy.' The little man walked after Nick and Sam, the cook, back into the kitchen. The door shut after them. The man called Max sat at the corner opposite George. He didn't look at George but looked in the mirror that ran along back of the counter. Henry's had been made over from a saloon into a lunch-counter.

'Well, bright boy,' Max said, looking into the mirror, 'why don't you say something?'

'What's it all about?'

'Hey, Al,' Max called, 'bright boy wants to know what it's all about.'

'Why don't you tell him?' Al's voice came from the kitchen.

'What do you think it's all about?'

'I don't know.'

'What do you think?'

Max looked into the mirror all the time he was talking.

'I wouldn't say.'

'Hey, Al, bright boy says he wouldn't say what he thinks it's all about.'

'I can hear you, all right,' Al said from the kitchen. He had propped open the slit that dishes passed through into the kitchen with a catsup bottle. 'Listen, bright boy,' he said from the kitchen to George. 'Stand a little farther along the bar. You move a little to the left, Max.' He was like a photographer arranging for a group picture.

'Talk to me, bright boy,' Max said. 'What do you think's going to happen?'

George did not say anything.

'I'll tell you,' Max said. 'We're going to kill a Swede. Do you know a big Swede named Ole Andreson?'

'Yes.'

'He comes here to eat every night, don't he?'

'Sometimes he comes here.'

'He comes here at six o'clock, don't he?'

'If he comes.'

'We know all that, bright boy,' Max said. 'Talk about something else. Ever go to the movies?'

'Once in a while.'

'You ought to go to the movies more. The movies are fine for a bright boy like you.'

'What are you going to kill Ole Andreson for? What did he ever do to you?'

'He never had a chance to do anything to us. He never even seen us.'

'And he's only going to see us once,' Al said from the kitchen.

'What are you going to kill him for, then?' George asked.

'We're killing him for a friend. Just to oblige a friend, bright boy.'

'Shut up,' said Al from the kitchen. 'You talk too goddam much.'

'Well, I got to keep bright boy amused. Don't I, bright boy?'

'You talk too damn much,' Al said. 'The nigger and my bright boy are amused by themselves. I got them tied up like a couple of girl friends in the convent.'

'I suppose you were in a convent?'

'You never know.'

'You were in a kosher convent. That's where you were.'

George looked up at the clock.

'If anybody comes in you tell them the cook is off, and if they keep after it, you tell them you'll go back and cook yourself. Do you get that, bright boy?'

'All right,' George said. 'What you going to do with us afterward?'

'That'll depend,' Max said. 'That's one of those things you never know at the time.'

George looked up at the clock. It was a quarter past six.

The door from the street opened. A street-car motorman came in.

'Hello, George,' he said. 'Can I get supper?'

'Sam's gone out,' George said. 'He'll be back in about half an hour.'

'I'd better go up the street,' the motorman said. George looked at the clock. It was twenty minutes past six.

'That was nice, bright boy,' Max said. 'You're a regular little gentleman.'

'He knew I'd blow his head off,' Al said from the kitchen.

'No,' said Max. 'It ain't that. Bright boy is nice. He's a nice boy. I like him.'

At six-fifty-five George said, 'He's not coming.'

Two other people had been in the lunch-room. Once George had gone out to the kitchen and made a ham-and-egg sandwich 'to go' that a man wanted to take with him. Inside the kitchen he saw Al, his derby hat tipped back, sitting on a stool beside the wicket with the muzzle of a sawed-off shotgun resting on the ledge. Nick and the cook were back to back in the corner, a towel tied in each of their mouths. George had cooked the sandwich, wrapped it up in oiled paper, put it in a bag, brought it in, and the man had paid for it and gone out.

'Bright boy can do everything,' Max said. 'He can cook and everything. You'd make some girl a nice wife, bright boy.'

'Yes?' George said. 'Your friend, Ole Andreson, isn't going to come.'

'We'll give him ten minutes,' Max said.

Max watched the mirror and the clock. The hands of the clock marked seven o'clock, and then five minutes past seven.

'Come on, Al,' said Max. 'We better go. He's not coming.'

'Better give him five minutes,' Al said from the kitchen.

In the five minutes a man came in, and George explained that the cook was sick.

'Why the hell don't you get another cook?' the man asked. 'Aren't you running a lunch-counter?' He went out.

'Come on, Al,' Max said.

'What about the two bright boys and the nigger?'

'They're all right.'

'You think so?'

'Sure. We're through with it.'

'I don't like it,' said Al. 'It's sloppy. You talk too much.'

'Oh, what the hell,' said Max. 'We got to keep amused, haven't we?'

'You talk too much, all the same,' Al said. He came out from the kitchen. The cut-off barrels of the shotgun made a slight bulge under the waist of his too tight-fitting overcoat. He straightened his coat with his gloved hands.

'So long, bright boy,' he said to George. 'You got a lot of luck.'

'That's the truth,' Max said. 'You ought to play the races, bright boy.'

The two of them went out the door. George watched them, through the window, pass under the arc-light and cross the street. In their tight overcoats and derby hats they looked like a vaudeville team. George went back through the swinging-door into the kitchen and untied Nick and the cook.

'I don't want any more of that,' said Sam, the cook. 'I don't want any more of that.'

Nick stood up. He had never had a towel in his mouth before.

'Say,' he said. 'What the hell?' He was trying to swagger it off.

'They were going to kill Ole Andreson,' George said. 'They were going to shoot him when he came in to eat.'

'Ole Andreson?'

'Sure.'

The cook felt the corners of his mouth with his thumbs. 'They all gone?' he asked.

'Yeah,' said George. 'They're gone now.'

'I don't like it,' said the cook. 'I don't like any of it at all.'

'Listen,' George said to Nick. 'You better go see Ole Andreson.'

'All right.'

'You better not have anything to do with it at all,' Sam, the cook, said. 'You better stay way out of it.'

'Don't go if you don't want to,' George said.

'Mixing up in this ain't going to get you anywhere,' the cook said. 'You stay out of it.'

'I'll go see him,' Nick said to George. 'Where does he live?'

The cook turned away.

'Little boys always know what they want to do,' he said.

'He lives up at Hirsch's rooming-house,' George said to Nick.

'I'll go up there.'

Outside the arc-light shone through the bare branches of a tree. Nick walked up the street beside the car-tracks and turned at the next arc-light down a side-street. Three houses up the street was Hirsch's rooming-house. Nick walked up the two steps and pushed the bell. A woman came to the door.

'Is Ole Andreson here?'

'Do you want to see him?'

'Yes, if he's in.'

Nick followed the woman up a flight of stairs and back to the end of the corridor. She knocked on the door.

'Who is it?'

'It's somebody to see you, Mr. Andreson,' the woman said. 'It's Nick Adams.'

'Come in.'

Nick opened the door and went into the room. Ole Andreson was lying on the bed with all his clothes on.

He had been a heavy-weight prize-fighter and he was too long for the bed. He lay with his head on two pillows. He did not look at Nick.

'What was it?' he asked.

'I was up at Henry's,' Nick said, 'and two fellows came in and tied up me and the cook, and they said they were going to kill you.'

It sounded silly when he said it. Ole Andreson said nothing.

'They put us out in the kitchen,' Nick went on. 'They were going to shoot you when you come in to supper.'

Ole Andreson looked at the wall and did not say anything.

'George thought I better come and tell you about it.'

'There isn't anything I can do about it,' Ole Andreson said.

'I'll tell you what they were like.'

'I don't want to know what they were like,' Ole Andreson said. He looked at the wall. 'Thanks for coming to tell me about it.'

'That's all right.'

Nick looked at the big man lying on the bed.

'Don't you want me to go and see the police?'

'No,' Ole Andreson said. 'That wouldn't do any good.'

'Isn't there something I could do?'

'No. There ain't anything to do.'

'Maybe it was just a bluff.'

'No. It ain't just a bluff.'

Ole Andreson rolled over toward the wall.

'The only thing is,' he said, talking toward the wall, 'I just can't make up my mind to go out. I been in here all day.'

'Couldn't you get out of town?'

'No,' Ole Andreson said. 'I'm through with all that running around.'

He looked at the wall.

'There ain't anything to do now.'

'Couldn't you fix it up some way?'

'No. I got in wrong.' He talked in the same flat voice. 'There ain't anything to do. After a while I'll make up my mind to go out.'

'I better go back and see George,' Nick said.

'So long,' said Ole Andreson. He did not look toward Nick. 'Thanks for coming around.'

Nick went out. As he shut the door he saw Ole Andreson with all his clothes on, lying on the bed looking at the wall.

'He's been in his room all day,' the landlady said downstairs. 'I guess he don't feel well. I said to him: "Mr. Andreson, you ought to go out and take a walk on a nice fall day like this," but he didn't feel like it.'

'He doesn't want to go out.'

'I'm sorry he don't feel well,' the woman said. 'He's an awfully nice man. He was in the ring, you know.'

'I know it.'

'You'd never know it except from the way his face is,' the woman said. They stood talking just inside the street door. 'He's just as gentle.'

'Well, good night, Mrs. Hirsch,' Nick said.

'I'm not Mrs. Hirsch,' the woman said. 'She owns the place. I just look after it for her. I'm Mrs. Bell.'

'Well, good night, Mrs. Bell,' Nick said.

'Good night,' the woman said.

Nick walked up the dark street to the corner under the arc-light, and then along the car-tracks to Henry's eating-house. George was inside, back of the counter.

'Did you see Ole?'

'Yes,' said Nick. 'He's in his room and he won't go out.'

The cook opened the door from the kitchen when he heard Nick's voice.

'I don't even listen to it,' he said and shut the door.

'Did you tell him about it?' George asked.

'Sure. I told him but he knows what it's all about.'

'What's he going to do?'

'Nothing.'

'They'll kill him.'

'I guess they will.'

'He must have got mixed up in something in Chicago.'

'I guess so,' said Nick.

'It's a hell of a thing.'

'It's an awful thing,' Nick said.

They did not say anything. George reached down for a towel and wiped the counter.

'I wonder what he did?' Nick said.

'Double-crossed somebody. That's what they kill them for.'

'I'm going to get out of this town,' Nick said.

'Yes,' said George, 'That's a good thing to do.'

'I can't stand to think about him waiting in the room and knowing he's going to get it. It's too damned awful.'

'Well,' said George, 'you better not think about it.'

A DAY'S WAIT

He came into the room to shut the windows while we were still in bed and I saw he looked ill. He was shivering, his face was white, and he walked slowly as though it ached to move.

'What's the matter, Schatz?'

'I've got a headache.'

'You better go back to bed.'

'No, I'm all right.'

'You go to bed. I'll see you when I'm dressed.'

But when I came downstairs he was dressed, sitting by the fire, looking a very sick and miserable boy of nine years. When I put my hand on his forehead I knew he had a fever.

'You go up to bed,' I said, 'you're sick.'

'I'm all right,' he said.

When the doctor came he took the boy's temperature.

'What is it?' I asked him.

'One hundred and two.'

Downstairs, the doctor left three different medicines in different coloured capsules with instructions for giving them. One was to bring down the fever, another a purgative, the third to overcome an acid condition. The germs of influenza can only exist in an acid condition, he explained. He seemed to know all about influenza and said there was nothing to worry about if the fever did not go above one hundred and four degress. This was a light epidemic of 'flu and there was no danger if you avoided pneumonia.

Back in the room I wrote the boy's temperature down and made a note of the time to give the various capsules.

'Do you want me to read to you?'

'All right. If you want to,' said the boy. His face was very white and there were dark areas under his eyes. He lay still in the bed and seemed very detached from what was going on.

I read aloud from Howard Pyle's *Book of Pirates;* but I could see he was not following what I was reading.

'How do you feel, Schatz?' I asked him.

'Just the same, so far,' he said.

I sat at the foot of the bed and read to myself while I waited for it to be time to give another capsule. It would have been natural for him to go to sleep, but when I looked up he was looking at the foot of the bed, looking very strangely.

'Why don't you try to go to sleep? I'll wake you up for the medicine.'

'I'd rather stay awake.'

After a while he said to me, 'You don't have to stay in here with me, Papa, if it bothers you.'

'It doesn't bother me.'

'No, I mean you don't have to stay if it's going to bother you.'

I thought perhaps he was a little light-headed, and after giving him the prescribed capsules at eleven o'clock I went out for a while.

It was a bright, cold day, the ground covered with a sleet that had frozen so that it seemed as if all the bare trees, the bushes, the cut brush and all the grass and the bare ground had been varnished with ice. I took the young Irish setter for a little walk up the road and along a frozen creek, but it was difficult to stand or walk on the glassy surface and the red dog slipped and slithered and I fell twice, hard, once dropping my gun and having it slide away over the ice.

We flushed a covey of quail under a high clay bank with overhanging brush and I killed two as they went out of sight over the top of the bank. Some of the covey lit in trees, but most of them scattered into brush piles and it was necessary to jump on the ice-coated mounds of brush several times before they would flush. Coming out while you were poised unsteadily on the icy, springy brush they made difficult shooting and I killed two, missed five, and started back pleased to have found a covey close to the house and happy there were so many left to find on another day.

At the house they said the boy had refused to let anyone come into the room.

'You can't come in,' he said. 'You mustn't get what I have.'

I went up to him and found him in exactly the position I had left him, white-faced, but with the tops of his cheeks flushed by the fever, staring still, as he had stared, at the foot of the bed.

I took his temperature.

'What is it?'

'Something like a hundred,' I said. It was one hundred and two and four-tenths.

'It was a hundred and two,' he said.

'Who said so?'

'The doctor.'

'Your temperature is all right,' I said. 'It's nothing to worry about.'

'I don't worry,' he said, 'but I can't keep from thinking.'

'Don't think,' I said. 'Just take it easy.'

'I'm taking it easy,' he said, and looked straight ahead. He was evidently holding tight on to himself about something.

'Take this with water.'

'Do you think it will do any good?'

'Of course it will.'

I sat down and opened the *Pirate* book and commenced to read, but I could see he was not following, so I stopped.

'About what time do you think I'm going to die?' he asked.

'What?'

'About how long will it be before I die?'

'You aren't going to die. What's the matter with you?'

'Oh, yes, I am. I heard him say a hundred and two.'

'People don't die with a fever of one hundred and two. That's a silly way to talk.'

'I know they do. At school in France the boys told me you can't live with forty-four degrees. I've got a hundred and two.'

He had been waiting to die all day, ever since nine o'clock in the morning.

'You poor Schatz,' I said. 'Poor old Schatz. It's like miles and kilometres. You aren't going to die. That's a different thermometer. On that thermometer thirty-seven is normal. On this kind it's ninety-eight.'

'Are you sure?'

'Absolutely,' I said. 'It's like miles and kilometres. You know, like how many kilometres we make when we do seventy miles in the car?'

'Oh,' he said.

But his gaze at the foot of the bed relaxed slowly. The hold over himself relaxed too, finally, and the next day it was very slack and he cried very easily at little things that were of no importance.

THE BATTLER

Nick stood up. He was all right. He looked up the track at the lights of the caboose going out of sight around the curve. There was water on both sides of the track, then tamarack swamp.

He felt of his knee. The pants were torn and the skin was barked. His hands were scraped and there were sand and cinders driven up under his nails. He went over to the edge of the track down the little slope to the water and washed his hands. He washed them carefully in the cold water, getting the dirt out from the nails. He squatted down and bathed his knee.

That lousy crut of a brakeman. He would get him some day. He would know him again. That was a fine way to act.

'Come here, kid,' he said. 'I got something for you.'

He had fallen for it. What a lousy kid thing to have done.

They would never suck him in that way again.

'Come here, kid, I got something for you.' Then *wham* and he lit on his hands and knees beside the track.

Nick rubbed his eye. There was a big bump coming up. He would have a black eye, all right. It ached already. That son of a crutting brakeman.

He touched the bump over his eye with his fingers. Oh, well, it was only a black eye. That was all he had gotten out of it. Cheap at the price. He wished he could see it. Could not see it looking into the water, though. It was dark and he was a long way off from anywhere. He wiped his hands on his trousers and stood up, then climbed the embankment to the rails.

He started up the track. It was well ballasted and made easy walking, sand and gravel packed between the ties, solid walking. The smooth roadbed like a causeway went on ahead through the swamp. Nick walked along. He must get to somewhere.

Nick had swung on to the freight train when it slowed down for the yards outside of Walton Junction. The train, with Nick on it, had passed through Kalkaska as it started to get dark. Now he must be nearly to Mancelona. Three or four miles of swamp. He stepped along the track, walking so he kept on the ballast between the ties, the swamp ghostly in the rising mist. His eye ached and he was hungry. He kept on hiking, putting the miles of track back of him. The swamp was all the same on both sides of the track.

Ahead there was a bridge. Nick crossed it, his boots ringing hollow on the iron. Down below the water showed black between the slits of ties. Nick kicked a loose spike and it dropped into the water. Beyond the bridge were hills. It was high and dark on both sides of the track. Up the track Nick saw a fire.

He came up the track toward the fire carefully. It was off to one side of the track, below the railway embankment. He had only seen the light from it. The track came out through a cut and where the fire was burning the country opened out and fell away into woods. Nick dropped carefully down the embankment and cut into the woods to come up to the fire through the trees. It was a beechwood forest and the fallen beechnut burrs were under his shoes as he walked between the trees. The fire was bright now, just at the edge of the trees. There was a man sitting by it. Nick waited behind the tree and watched. The man looked to be alone. He was sitting there with his head in his hands looking at the fire. Nick stepped out and walked into the firelight.

The man sat there looking into the fire. When Nick stopped quite close to him he did not move.

'Hello!' Nick said.

The man looked up.

'Where did you get the shiner?' he said.

'A brakeman busted me.'

'Off the through freight?'

'Yes.'

'I saw the bastard,' the man said. 'He went through here 'bout an hour and a half ago. He was walking along the top of the cars slapping his arms and singing.'

'The bastard!'

'It must have made him feel good to bust you,' the man said seriously.

'I'll bust him.'

'Get him with a rock sometime when he's going through,' the man advised.

'I'll get him.'

'You're a tough one, aren't you?'

'No,' Nick answered.

'All you kids are tough.'

'You got to be tough,' Nick said.

'That's what I said.'

The man looked at Nick and smiled. In the firelight Nick saw that his face was misshapen. His nose was sunken, his eyes were slits, he had queer-shaped lips. Nick did not perceive all this at once, he only saw the man's face was queerly formed and mutilated. It was like putty in colour. Dead looking in the firelight.

'Don't you like my pan?' the man asked.

Nick was embarrassed.

'Sure,' he said.

'Look here!' the man took off his cap.

He had only one ear. It was thickened and tight against the side of his head. Where the other ear should have been there was a stump.

'Ever see one like that?'

'No,' said Nick. It made him a little sick.

'I could take it,' the man said. 'Don't you think I could take it, kid?'

'You bet!'

'They all bust their hands on me,' the little man said. 'They couldn't hurt me.'

He looked at Nick. 'Sit down,' he said. 'Want to eat?'

'Don't bother,' Nick said. 'I'm going on to the town.'

'Listen!' the man said. 'Call me Ad.'

'Sure!'

'Listen,' the little man said. 'I'm not quite right.'

'What's the matter?'

'I'm crazy.'

He put on his cap. Nick felt like laughing.

'You're all right,' he said.

'No, I'm not. I'm crazy. Listen, you ever been crazy?'

'No,' Nick said. 'How does it get you?'

'I don't know,' Ad said. 'When you got it you don't know about it. You know me, don't you?'

'No.'

'I'm Ad Francis.'

'Honest to God?'

'Don't you believe it?'

'Yes.'

Nick knew it must be true.

'You know how I beat them?'

'No,' Nick said.

'My heart's slow. It only beats forty a minute. Feel it.'

Nick hesitated.

'Come on,' the man took hold of his hand. 'Take hold of my wrist. Put your fingers there.'

The little man's wrist was thick and the muscles bulged above the bone. Nick felt the slow pumping under his fingers.

'Got a watch?'

'No.'

'Neither have I,' Ad said. 'It ain't any good if you haven't got a watch.'

Nick dropped his wrist.

'Listen,' Ad Francis said. 'Take a hold again. You count and I'll count up to sixty.'

Feeling the slow hard throb under his fingers Nick started to count. He heard the little man counting slowly, one, two, three, four, five, and on—aloud.

'Sixty,' Ad finished.' 'That's a minute. What did you make it?'

'Forty,' Nick said.

'That's right,' Ad said happily. 'She never speeds up.'

A man dropped down the railroad embankment and came across the clearing to the fire.

'Hello, Bugs!' Ad said.

'Hello!' Bugs answered. It was a negro's voice. Nick knew from the way he walked that he was a negro He stood with his back to them, bending over the fire. He straightened up.

'This is my pal Bugs,' Ad said. 'He's crazy, too.'

'Glad to meet you,' Bugs said. 'Where you say you're from?'

'Chicago,' Nick said.

'That's a fine town,' the negro said. 'I didn't catch your name.'

'Adams. Nick Adams.'

'He says he's never been crazy, Bugs,' Ad said.

'He's got a lot coming to him,' the negro said. He was unwrapping a package by the fire.

'When are we going to eat, Bugs?' the prize-fighter asked.

'Right away.'

'Are you hungry, Nick?'

'Hungry as hell.'

'Hear that, Bugs?'

'I hear most of what goes on.'

'That ain't what I asked you.'

'Yes. I heard what the gentleman said.'

Into a skillet he was laying slices of ham. As the skillet grew hot the grease sputtered and Bugs, crouching on long nigger legs over the fire, turned the ham and broke eggs into the skillet, tipping it from side to side to baste the eggs with the hot fat.

'Will you cut some bread out of that bag, Mister Adams?' Bugs turned from the fire.

'Sure.'

Nick reached in the bag and brought out a loaf of bread. He cut six slices. Ad watched him and leaned forward.

'Let me take your knife, Nick,' he said.

'No, you don't,' the negro said. 'Hang on to your knife, Mister Adams.'

The prize-fighter sat back.

'Will you bring me the bread, Mister Adams?' Bugs asked. Nick brought it over.

'Do you like to dip your bread in the ham fat?' the negro asked.

'You bet!'

'Perhaps we'd better wait until later. It's better at the finish of the meal. Here.'

The negro picked up a slice of ham and laid it on one of the pieces of bread, then slid an egg on top of it.

'Just close that sandwich, will you, please, and give it to Mister Francis.'

Ad took the sandwich and started eating.

'Watch out how that egg runs,' the negro warned. 'This is for you, Mister Adams. The remainder for myself.'

Nick bit into the sandwich. The negro was sitting opposite him beside Ad. The hot fried ham and eggs tasted wonderful.

'Mister Adams is right hungry,' the negro said. The little man whom Nick knew by name as a former champion fighter was silent. He had said nothing since the negro had spoken about the knife.

'May I offer you a slice of bread dipped right in the hot ham fat?' Bugs said.

'Thanks a lot.'

The little white man looked at Nick.

'Will you have some, Mister Adolph Francis?' Bugs offered from the skillet.

Ad did not answer. He was looking at Nick.

'Mister Francis?' came the nigger's soft voice.

Ad did not answer. He was looking at Nick.

'I spoke to you, Mister Francis,' the nigger said softly.

Ad kept on looking at Nick. He had his cap down over his eyes. Nick felt nervous.

'How the hell do you get that way?' came out from under the cap sharply at Nick. 'Who the hell do you think you are? You're a snotty bastard. You come in here where nobody asks you and eat a man's food and when he asks to borrow a knife you get snotty.'

He glared at Nick, his face was white and his eyes almost out of sight under the cap.

'You're a hot sketch. Who the hell asked you to butt in here?'

'Nobody.'

'You're damn right nobody did. Nobody asked you to stay either. You come in here and act snotty about my face and smoke my cigars and drink my liquor and then talk snotty. Where the hell do you think you get off?'

Nick said nothing. Ad stood up.

'I'll tell you, you yellow-livered Chicago bastard. You're going to get your can knocked off. Do you get that?'

Nick stepped back. The little man came towards him slowly, stepping flat-footed forward, his left foot stepping forward, his right dragging up to it.

'Hit me,' he moved his head. 'Try and hit me.'

'I don't want to hit you.'

'You won't get out of it that way. You're going to take a beating, see? Come on and lead at me.'

'Cut it out,' Nick said.

'All right, then, you bastard.'

The little man looked down at Nick's feet. As he looked down, the negro, who had followed behind him as he moved away from the fire, set himself and tapped him across the base of the skull. He fell forward and Bugs dropped the cloth-wrapped blackjack on the grass. The little man lay there, his face in the grass. The negro picked him up, his head hanging, and carried him to the fire. His face looked bad, the eyes open. Bugs laid him down gently.

'Will you bring me the water in the bucket, Mister Adams?' he said. 'I'm afraid I hit him just a little hard.'

The negro splashed water with his hand on the man's face and pulled his ear gently. The eyes closed.

Bugs stood up.

'He's all right,' he said. 'There's nothing to worry about. I'm sorry, Mister Adams.'

'It's all right.' Nick was looking down at the little man. He saw the blackjack on the grass and picked it up. It had a flexible handle and was limber in his hand. It was made of worn black leather with a handkerchief wrapped around the heavy end.

'That's a whalebone handle,' the negro smiled. 'They don't make them any more. I didn't know how well you could take care of yourself and, anyway, I didn't want you to hurt him or mark him up no more than he is.'

The negro smiled again.

'You hurt him yourself.'

'I know how to do it. He won't remember nothing of it. I have to do it to change him when he gets that way.'

Nick was still looking down at the little man, lying, his eyes closed in the firelight. Bugs put some wood on the fire.

'Don't you worry about him none, Mister Adams. I seen him like this plenty of times before.'

'What made him crazy?' Nick asked.

'Oh, a lot of things,' the negro answered from the fire. 'Would you like a cup of this coffee, Mister Adams?'

He handed Nick the cup and smoothed the coat he had placed under the unconscious man's head.

'He took too many beatings, for one thing,' the negro sipped the coffee. 'But that just made him sort of simple. Then his sister was his manager and they was always being written up in the papers all about brothers and sisters and how she loved her brother and how he loved his sister, and then they got married in New York and that made a lot of unpleasantness.'

'I remember about it.'

'Sure. Of course they wasn't brother and sister no more than a rabbit, but there was a lot of people didn't like it either way and they commenced to have disagreements, and one day she just went off and never come back.'

He drank the coffee and wiped his lips with the pink palm of his hand.

'He just went crazy. Will you have some more coffee, Mister Adams?'

'Thanks.'

'I seen her a couple of times,' the negro went on. 'She was an awful good-looking woman. Looked enough like him to be twins. He wouldn't be bad-looking without his face all busted.'

He stopped. The story seemed to be over.

'Where did you meet him?' asked Nick.

'I met him in jail,' the negro said. 'He was busting people all the time after she went away and they put him in jail. I was in for cuttin' a man.'

He smiled, and went on soft-voiced:

'Right away I liked him and when I got out I looked him up. He likes to think I'm crazy and I don't mind. I like to be with him and I like seeing the country and I don't have to commit no larceny to do it. I like living like a gentleman.'

'What do you all do?' Nick asked.

Oh, nothing Just move around. He's got money.'

'He must have made a lot of money.'

'Sure. He spent all his money, though. Or they took it away from him. She sends him money.'

He poked up the fire.

'She's a mighty fine woman,' he said. 'She looks enough like him to be his own twin.'

The negro looked over at the little man, lying breathing heavily. His blond hair was down over his forehead. His mutilated face looked childish in repose.

'I can wake him up any time now, Mister Adams. If you don't mind I wish you'd sort of pull out. I don't like to not be hospitable, but it might disturb him back again to see you. I hate to have to thump him and it's the only thing to do when he gets started. I have to sort of keep him away from people. You don't mind, do you, Mister Adams? No, don't thank me, Mister Adams. I'd have warned you about him but he seemed to have taken such a liking to you and I thought things were going to be all right. You'll hit a town about two miles up the track. Mancelona they call it. Goodbye. I wish we could ask you to stay the night but it's just out of the question. Would you like to take some of that ham and some bread with you? No? You better take a sandwich,' all this in a low, smooth, polite nigger voice.

'Good. Well, goodbye, Mister Adams. Goodbye and good luck!'

Nick walked away from the fire across the clearing to the railway tracks. Out of range of the fire he listened. The low soft voice of the negro was talking. Nick could not hear the words. Then he heard the little man say, 'I got an awful headache, Bugs.'

'You'll feel better, Mister Francis,' the negro's voice soothed. 'Just you drink a cup of this hot coffee.'

Nick climbed the embankment and started up the track. He found he had a ham sandwich in his hand and put it in his pocket. Looking back from the mounting grade before the track curved into the hills he could see the firelight in the clearing.

AFTER THE STORM

It wasn't about anything, something about making punch, and then we started fighting and I slipped and he had me down kneeling on my chest and choking me with both hands like he was trying to kill me and all the time I was trying to get the knife out of my pocket to cut him loose. Everybody was too drunk to pull him off me. He was choking me and hammering my head on the floor and I got the knife out and opened it up; and I cut the muscle right across his arm and he let go of me. He couldn't have held on if he wanted to. Then he rolled and hung on to that arm and started to cry and I said:

'What the hell you want to choke me for?'

I'd have killed him. I couldn't swallow for a week. He hurt my throat bad.

Well, I went out of there and there were plenty of them with him and some came out after me and I made a turn and was down by the docks and I met a fellow and he said somebody killed a man up the street. I said, 'Who killed him?' and he said, 'I don't know who killed him but he's dead all right,' and it was dark and there was water standing in the street and no lights and windows broke, and boats all up in the town and trees blown down and everything all blown and I got a skiff and went out and found my boat where I had her inside of Mango Key and she was all right only she was full of water. So I bailed her out and pumped her out and there was a moon but plenty of clouds and still plenty rough and I took it down along; and when it was daylight I was off Eastern Harbour.

Brother, that was some storm. I was the first boat out and you never saw water like that was. It was just as white as a lye barrel and coming from Eastern Harbour to Sou'west Key you couldn't recognize the shore. There was a big channel blown right out through the middle of the beach. Trees and all blown out and a channel cut through and all the water white as chalk and everything on it; branches and whole trees and dead birds, and all floating.

Inside the keys were all the pelicans in the world and all kinds of birds flying. They must have gone inside there when they knew it was coming.

I lay at Sou'west Key a day and nobody came after me. I was the first boat out and I seen a spar floating and I knew there must be a wreck and I started out to look for her. I found her. She was a three-masted schooner and I could just see the stumps of her spars out of water. She was in too deep water and I didn't get anything off of her. So I went on looking for something else. I had the start on all of them and I knew I ought to get whatever there was. I went on down over the sand-bar from where I left that three-masted schooner and I didn't find anything and I went on a long way. I was way out toward the quicksands and I didn't find anything so I went on. Then when I was in sight of the Rebecca Light I saw all kinds of birds making over something and I headed over for them to see what it was and there was a cloud of birds all right.

I could see something looked like a spar up out of the water and when I got over close the birds all went up in the air and stayed all around me. The water was clear out there and there was a spar of some kind sticking out just above the water and when I come up close to it I saw it was all dark under water like a long shadow and I came right over it and there under water was a liner; just lying there all under water as big as the whole world. I drifted over her in the boat. She lay on her side and the stern was deep down. The port holes were all shut tight and I could see the glass shine in the water and the whole of her; the biggest boat I ever saw in my life laying there and I went along the whole length of her and then I went over and anchored and I had the skiff on the deck forward and I shoved it down into the water and sculled over with the birds all around me.

I had a water glass like we use sponging and my hand shook so I could hardly hold it. All the port holes were shut that you could see going along over her but way down below near the bottom something must have been open because there were pieces of things floating out all the time. You couldn't tell what they were. Just pieces. That's what the birds were after. You never saw so many birds. They were all around me; crazy yelling.

I could see everything sharp and clear. I could see her rounded over and she looked a mile long under the water. She was lying on a clear white bank of sand and the spar was a sort of foremast or some sort of tackle that slanted out of water the way she was laying on her side. Her bow wasn't very far under. I could stand on the letters of her name on her bow and my head was just out of water. But the nearest port hole was twelve feet down. I could just reach it with the grains pole and I tried to break it with that but I couldn't. The glass was too stout. So I sculled back to the boat and got a wrench and lashed it to the end of the grains pole and I couldn't break it. There I was looking down through the glass at that liner with everything in her and I was the first one to her and I couldn't get into her. She must have had five million dollars' worth in her.

It made me shaky to think how much she must have in her. Inside the port hole that was closed I could see something but I couldn't make it out through the water glass. I couldn't do any good with the grains pole and I took off my clothes and stood and took a couple of deep breaths and dove over off the stern with the wrench in my hand and swam down. I could hold on for a second to the edge of the port hole and I could see in and there was a woman inside with her hair floating all out. I could see her floating plain and I hit the glass twice with the wrench hard and I heard the noise clink in my ears but it wouldn't break and I had to come up.

I hung on to the dinghy and got my breath and then I climbed in and took a couple of breaths and dove again. I swam down and took hold of the edge of the port hole with my fingers and held it and hit the glass as hard as I could with the wrench. I could see the woman floated in the water through the glass. Her hair was tied once close to her head and it floated all out in the water. I could see the rings on one of her hands. She was right up close to the port hole and I hit the glass twice and I didn't even crack it. When I came up I thought I wouldn't make it to the top before I'd have to breathe.

I went down once more and I cracked the glass, only cracked it, and when I came up my nose was bleeding and I stood on the bow of the liner with my bare feet on the letters of her name and my head just out and rested there and then I swam over to the skiff and pulled up into it and sat there waiting for my head to stop aching and looking down into the water glass, but I bled so I had to wash out the water glass. Then I lay back in the skiff and held my hand under my nose to stop it and I lay there with my head back looking up and there was a million birds above and all around.

When I quit bleeding I took another look through the glass and then I sculled over to the boat to try and find something heavier than the wrench but I couldn't find a thing; not even a sponge hook. I went back and the water was clearer all the time and you could see everything that floated out over that white bank of sand. I looked for sharks but there weren't any. You could have seen a shark a long way away. The water was so clear and the sand white. There was a grapple for an anchor on the skiff and I cut it off and went overboard and down with it. It carried me right down and past the port hole and I grabbed and couldn't hold anything and went on down and down, sliding along the curved side of her. I had to let go of the grapple. I heard it bump once and it seemed like a year before I came up through to the top of the water. The skiff was floated away with the tide and I swam over to her with my nose bleeding in the water while I swam and I was plenty glad there weren't sharks; but I was tired.

My head felt cracked open and I lay in the skiff and rested and then I sculled back. It was getting along in the afternoon. I went down once more with the wrench and it didn't do any good. That wrench was too light. It wasn't any good diving unless you had a big hammer or something heavy enough to do good. Then I lashed the wrench to the grains pole again and I watched through the water glass and pounded on the glass and hammered until the wrench came off and I saw it in the glass, clear and sharp, go sliding down along her and then off and down to the quicksand and go in. Then I couldn't do a thing. The wrench was gone and I'd lost the grapple so I sculled back to the boat. I was too tired to get the skiff aboard and the sun was pretty low. The birds were all pulling out and leaving her and I headed for Sou'west Key towing the skiff and the birds going on ahead of me and behind me. I was plenty tired.

That night it came on to blow and it blew for a week. You couldn't get out to her. They come out from town and told me the fellow I'd had to cut was all right except for his arm and I went back to town and they put me under five hundred dollar bond. It came out all right because some of them, friends of mine, swore he was after me with an axe, but the time we got back out to her the Greeks had blown her open and cleaned her out. They got the safe out with dynamite. Nobody ever knows how much they got. She carried gold and they got it all. They stripped her clean. I found her and I never got a nickel out of her.

It was a hell of a thing all right. They say she was just outside of Havana harbour when the hurricane hit and she couldn't get in or the owners wouldn't let the captain chance coming in; they say he wanted to try; so she had to go with it and in the dark they were running with it trying to go through the gulf between Rebecca and Tortugas when she struck on the quicksands. Maybe her rudder was carried away. Maybe they weren't even steering. But anyway they couldn't have known they were quicksands and when she struck the captain must have ordered them to open up the ballast tanks so she'd lay solid. But it was quicksand she'd hit, and when they opened the tank she went in stern first and then over on her beam ends. There were four hundred and fifty passengers and the crew on board of her, and they must all have been aboard of her when I found her. They must have opened the tanks as soon as she struck, and the minute she settled on it the quicksands took her down. Then her boilers must have burst, and that must have been what made those pieces that came out. It was funny there weren't any sharks though. There wasn't a fish. I could have seen them on that clear white sand.

Plenty of fish now though; jewfish, the biggest kind. The biggest part of her's under the sand now but they live inside of her; the biggest kind of jewfish. Some weigh three to four hundred pounds. Sometime we'll go out and get some. You can see the Rebecca light from where she is. They've got a buoy on her now. She's right at the end of the quicksand right at the edge of the gulf. She only missed going through by about a hundred yards. In the dark in the storm they just missed it; raining the way it was they couldn't have seen the Rebecca. Then they're not used to that sort of thing. The captain of a liner isn't used to scudding that way. They have a course, and they tell me they set some sort of a compass and it steers itself. They probably didn't know where they were when they ran with that blow but they come close to making it. Maybe they'd lost the rudder though. Anyway there wasn't another thing for them to hit till they'd get to Mexico once they were in that gulf. Must have been something though when they struck in that rain and wind and he told them to open her tanks. Nobody could have been on deck in that blow and rain. Everybody must have been below. They couldn't have lived on deck. There must have been some scenes inside all right because you know she settled fast. I saw that wrench go into the sand. The captain couldn't have known it was quicksand when she struck unless he knew these waters. He just knew it wasn't rock. He must have seen it all up in the bridge. He must have known what it was about when she settled. I wonder how fast she made it. I wonder if the mate was there with him. Do you think they stayed inside the bridge or do you think they took it outside? They never found any bodies. Not a one. Nobody floating. They float a long way with lifebelts too. They must have took it inside. Well, the Greeks got it all. Everything. They must have come fast all right. They picked her clean. First there was the birds, then me, then the Greeks, and even the birds got more out of her than I did.

THE DOCTOR AND THE DOCTOR'S WIFE

Dick Boulton came from the Indian camp to cut up logs for Nick's father. He brought his son Eddy and another Indian named Billy Tabeshaw with him. They came in through the back gate out of the woods, Eddy carrying the long cross-cut saw. It flopped over his shoulder and made a musical sound as he walked. Billy Tabeshaw carried two big cant-hooks. Dick had three axes under his arm.

He turned and shut the gate. The others went on ahead of him down to the lake shore where the logs were buried in the sand.

The logs had been lost from the big log booms that were towed down the lake to the mill by the steamer *Magic*. They had drifted up on to the beach and if nothing were done about them sooner or later the crew of the *Magic* would come along the shore in a rowboat, spot the logs, drive an iron spike with a ring on it into the end of each one and then tow them out into the lake to make a new boom. But the lumbermen might never come for them because a few logs were not worth the price of a crew to gather them. If no one came for them they would be left to waterlog and rot on the beach.

Nick's father always assumed that this was what would happen, and hired the Indians to come down from the camp and cut the logs up with the cross-cut saw and split them with a wedge to make cord wood and chunks for the open fireplace. Dick Boulton walked around past the cottage down to the lake. There were four big beech logs lying almost buried in the sand. Eddy hung the saw up by one of its handles in the crotch of a tree. Dick put the three axes down on the little dock. Dick was a half-breed and many of the farmers around the lake believed he was really a white man. He was very lazy but a great worker once he was started. He took a plug of tobacco out of his pocket, bit off a chew and spoke in Ojibway to Eddy and Billy Tabeshaw.

They sunk the ends of their cant-hooks into one of the logs and swung against it to loosen it in the sand. They swung their weight against the shafts of the cant-hooks. The log moved in the sand. Dick Boulton turned to Nick's father.

'Well, Doc,' he said, 'that's a nice lot of timber you've stolen.'

'Don't talk that way, Dick,' the doctor said. 'It's driftwood.'

Eddy and Billy Tabeshaw had rocked the log out of the wet sand and rolled it towards the water.

'Put it right in,' Dick Boulton shouted.

'What are you doing that for?' asked the doctor.

'Wash it off. Clean off the sand on account of the saw. I want to see who it belongs to,' Dick said.

The log was just awash in the lake. Eddy and Billy Tabeshaw leaned on their cant-hooks sweating in the sun. Dick kneeled down in the sand and looked at the mark of the scaler's hammer in the wood at the end of the log.

'It belongs to White and McNally,' he said, standing up and brushing off his trousers knees.

The doctor was very uncomfortable.

'You'd better not saw it up then, Dick,' he said, shortly.

'Don't get huffy, Doc,' said Dick. 'Don't get huffy. I don't care who you steal from. It's none of my business.'

'If you think the logs are stolen, leave them alone and take your tools back to the camp,' the doctor said. His face was red.

'Don't go off at half-cock, Doc,' Dick said. He spat tobacco juice on the log. It slid off, thinning in the water. 'You know they're stolen as well as I do. It don't make any difference to me.'

'All right. If you think the logs are stolen, take your stuff and get out.'

'Now, Doc—'

'Take your stuff and get out.'

'Listen, Doc.'

'If you call me Doc once again, I'll knock your eye teeth down your throat.'

'Oh, no, you won't, Doc.'

Dick Boulton looked at the doctor. Dick was a big man. He knew how big a man he was. He liked to get into fights. He was happy. Eddy and Billy Tabeshaw leaned on their cant-hooks and looked at the doctor. The doctor chewed the beard on his lower lip and looked at Dick Boulton. Then he turned away and walked up the hill to the cottage. They could see from his back how angry he was. They all watched him walk up the hill and go inside the cottage.

Dick said something in Ojibway. Eddy laughed but Billy Tabeshaw looked very serious. He did not understand English but he had sweat all the time the row was going on. He was fat with only a few hairs of moustache like a Chinaman. He picked up the two cant-hooks. Dick picked up the axes and Eddy took the saw down from the tree. They started off and walked up past the cottage and out the back gate into the woods. Dick left the gate open. Billy Tabeshaw went back and fastened it. They were gone through the woods.

In the cottage the doctor, sitting on the bed in his room, saw a pile of medical journals on the floor by the bureau. They were still in their wrappers unopened. It irritated him.

'Aren't you going back to work, dear?' asked the doctor's wife from the room where she was lying with the blinds drawn.

'No!'

'Was anything the matter?'

'I had a row with Dick Boulton.'

'Oh,' said his wife. 'I hope you didn't lose your temper, Henry.'

'No,' said the doctor.

'Remember, that he who ruleth his spirit is greater than he that taketh a city,' said his wife. She was a Christian

Scientist. Her Bible, her copy of *Science and Health* and her *Quarterly* were on a table beside her bed in the darkened room.

Her husband did not answer. He was sitting on his bed now, cleaning a shotgun. He pushed the magazine full of the heavy yellow shells and pumped them out again. They were scattered on the bed.

'Henry,' his wife called. Then paused a moment. 'Henry!'

'Yes,' the doctor said.

'You didn't say anything to Boulton to anger him, did you?'

'No,' said the doctor.

'What was the trouble about, dear?'

'Nothing much.'

'Tell me, Henry. Please don't try and keep anything from me. What was the trouble about?'

'Well, Dick owes me a lot of money for pulling his squaw through pneumonia and I guess he wanted a row so he wouldn't have to take it out in work.'

His wife was silent. The doctor wiped his gun carefully with a rag. He pushed the shells back in against the spring of the magazine. He sat with the gun on his knees. He was very fond of it. Then he heard his wife's voice from the darkened room.

'Dear, I don't think, I really don't think that anyone would really do a thing like that.'

'No?' the doctor said.

'No. I can't really believe that anyone would do a thing of that sort intentionally.'

The doctor stood up and put the shotgun in the corner behind the dresser.

'Are you going out, dear?' his wife said.

'I think I'll go for a walk,' the doctor said.

'If you see Nick, dear, will you tell him his mother wants to see him?' his wife said.

The doctor went out on the porch. The screen door slammed behind him. He heard his wife catch her breath when the door slammed.

'Sorry,' he said, outside her window with the blinds drawn.

'It's all right, dear,' she said.

He walked in the heat out the gate and along the path into the hemlock woods. It was cool in the woods even on such a hot day. He found Nick sitting with his back against a tree, reading.

'Your mother wants you to come and see her,' the doctor said.

'I want to go with you,' Nick said.

His father looked down at him.

'All right. Come on, then,' his father said. 'Give me the book, I'll put it in my pocket.'

'I know where there's black squirrels, Daddy,' Nick said.

'All right,' said his father. 'Let's go there.'

OLD MAN AT THE BRIDGE

An old man with steel-rimmed spectacles and very dusty clothes sat by the side of the road. There was a pontoon bridge across the river and carts, trucks, and men, women and children were crossing it. The muledrawn carts staggered up the steep bank from the bridge with soldiers helping push against the spokes of the wheels. The trucks ground up and away heading out of it all and the peasants plodded along in the ankle-deep dust. But the old man sat there without moving. He was too tired to go any farther.

It was my business to cross the bridge, explore the bridgehead beyond and find out to what point the enemy had advanced. I did this and returned over the bridge. There were not so many carts now and very few people on foot, but the old man was still there.

'Where do you come from?' I asked him.

'From San Carlos,' he said, and smiled.

That was his native town and so it gave him pleasure to mention it and he smiled.

'I was taking care of animals,' he explained.

'Oh,' I said, not quite understanding.

'Yes,' he said, 'I stayed, you see, taking care of animals. I was the last one to leave the town of San Carlos.'

He did not look like a shepherd nor a herdsman and I looked at his black dusty clothes and his grey dusty face and his steel-rimmed spectacles and said, 'What animals were they?'

'Various animals,' he said, and shook his head. 'I had to leave them.'

I was watching the bridge and the African-looking country of the Ebro Delta and wondering how long now it would be before we would see the enemy, and listening all the while for the first noises that would signal that ever mysterious event called contact, and the old man still sat there.

'What animals were they?' I asked.

'There were three animals altogether,' he explained. 'There were two goats and a cat and then there were four pairs of pigeons.'

'And you had to leave them?' I asked.

'Yes. Because of the artillery. The captain told me to go because of the artillery.'

'And you have no family?' I asked, watching the far end of the bridge where a few last carts were hurrying down the slope of the bank.

'No,' he said, 'only the animals I stated. The cat, of course, will be all right. A cat can look out for itself, but I cannot think what will become of the others.'

'What politics have you?' I asked.

'I am without politics,' he said. 'I am seventy-six years old. I have come twelve kilometres now and I think now I can go no farther.'

'This is not a good place to stop,' I said. 'If you can make it, here are trucks up the road where it forks for Tortosa.'

'I will wait a while,' he said, 'and then I will go. Where do the trucks go?'

'Towards Barcelona,' I told him.

'I know of no one in that direction,' he said, 'but thank you very much. Thank you again very much.'

He looked at me very blankly and tiredly, then said, having to share his worry with someone, 'The cat will be all right, I am sure. There is no need to be unquiet about the cat. But the others. Now what do you think about the others?'

'Why, they'll probobly come through it all right.'

'You think so?'

'Why not?' I said, watching the far bank where now there were no carts.

'But what will they do under the artillery when I was told to leave because of the artillery?'

'Did you leave the dove cage unlocked?' I asked.

'Yes.'

'Then they'll fly.'

'Yes, certainly they'll fly. But the others. It's better not to think about the others,' he said.

'If you are rested I would go,' I urged. 'Get up and try to walk now.'

'Thank you,' he said and got to his feet, swayed from side to side and then sat down backwards in the dust.

'I was taking care of animals,' he said dully, but no longer to me. 'I was only taking care of animals.'

There was nothing to do about him. It was Easter Sunday and the Fascists were advancing towards the Ebro. It was a grey overcast day with a low ceiling so their planes were not up. That and the fact that cats know how to look after themselves was all the good luck that old man would ever have.

THE UNDEFEATED

Manuel Garcia climbed the stairs to Don Miguel Retana's office. He set down his suitcase and knocked on the door. There was no answer. Manuel, standing in the hallway, felt there was someone in the room. He felt it through the door.

'Retana,' he said, listening.

There was no answer.

He's there, all right, Manuel thought.

'Retana,' he said and banged the door.

'Who's there?' said someone in the office.

'Me, Manolo,' Manuel said.

'What do you want?' asked the voice.

'I want to work,' Manuel said.

Something in the door clicked several times and it swung open. Manuel went in, carrying his suitcase.

A little man sat behind a desk at the far side of the room. Over his head was a bull's head, stuffed by a Madrid taxidermist; on the walls were framed photographs and bull-fight posters.

The little man sat looking at Manuel.

'I thought they'd killed you,' he said.

Manuel knocked with his knuckles on the desk. The little man sat looking at him across the desk.

'How many corridas you had this year?' Retana asked.

'One,' he answered.

'Just that one?' the little man asked.

'That's all.'

'I read about it in the papers,' Retana said. He leaned back in the chair and looked at Manuel.

Manuel looked up at the stuffed bull. He had seen it often before. He felt a certain family interest in it. It had killed his brother, the promising one, about nine years ago. Manuel remembered the day. There was a brass plate on the oak shield the bull's head was mounted on. Manuel could not read it, but he imagined it was in memory of his brother. Well, he had been a good kid.

The plate said: 'The Bull "Mariposa" of the Duke of Veragua, which accepted 9 varas for 7 caballos, and caused the death of Antonio Garcia, Novillero, April 27th, 1909.'

Retana saw him looking at the stuffed bull's head.

'The lot the Duke sent me for Sunday will make a scandal,' he said. 'They're all bad in the legs. What do they say about them at the Café?'

'I don't know,' Manuel said. 'I just got in.'

'Yes,' Retana said. 'You still have your bag.'

He looked at Manuel, leaning back behind the big desk.

'Sit down,' he said. 'Take off your cap.'

Manuel sat down; his cap off, his face was changed. He looked pale, and his coleta pinned forward on his head, so that it would not show under the cap, gave him a strange look.

'You don't look well,' Retana said.

'I just got out of the hospital,' Manuel said.

'I heard they'd cut your leg off,' Retana said.

'No,' said Manuel. 'It got all right.'

Retana leaned forward across the desk and pushed a wooden box of cigarettes toward Manuel.

'Have a cigarette,' he said.

'Thanks.'

Manuel lit it.

'Smoke?' he said, offering the match to Retana.

'No,' Retana waved his hand, 'I never smoke.'

Retana watched him smoking.

'Why don't you get a job and go to work?' he said.

'I don't want to work.' Manuel said. 'I am a bull-fighter.'

'There aren't any bull-fighters any more,' Retana said.

'I'm a bull-fighter,' Manuel said.

'Yes, while you're in there,' Retana said.

Manuel laughed.

Retana sat, saying nothing and looking at Manuel.

'I'll put you in a nocturnal if you want,' Retana offered.

'When?' Manuel asked.

'Tomorrow night.'

'I don't like to substitute for anybody,' Manuel said. That was the way they all got killed. That was the way Salvador got killed. He tapped with his knuckles on the table.

'It's all I've got,' Retana said.

'Why don't you put me on next week?' Manuel suggested.

'You wouldn't draw,' Retana said. 'All they want is Litri and Rubito and La Torre. Those kids are good.'

'They'd come to see me get it,' Manuel said, hopefully.

'No, they wouldn't. They don't know who you are any more.'

'I've got a lot of stuff,' Manuel said.

'I'm offering to put you on tomorrow night,' Retana said.

'You can work with young Hernandez and kill two novillos after the Charlots.'

'Whose novillos?' Manuel asked.

'I don't know. Whatever stuff they've got in the corrals. What the veterinaries won't pass in the daytime.'

'I don't like to substitute,' Manuel said.

'You can take it or leave it,' Retana said. He leaned forward over the papers. He was no longer interested. The appeal that Manuel had made to him for a moment when he thought of the old days was gone. He would like to get him to substitute for Larita because he could get him cheaply. He could get others cheaply too. He would like to help him, though. Still he had given him the chance. It was up to him.

'How much do I get?' Manuel asked. He was still playing with the idea of refusing. But he knew he could not refuse.

'Two hundred and fifty pesetas,' Retana said. He had thought of five hundred, but when he opened his mouth it said two hundred and fifty.

'You pay Villalta seven thousand,' Manuel said.

'You're not Villalta,' Retana said.

'I know it,' Manuel said.

'He draws it, Manolo,' Retana said in explanation.

'Sure,' said Manuel. He stood up. 'Give me three hundred, Retana.'

'All right,' Retana agreed. He reached in the drawer for a paper.

'Can I have fifty now?' Manuel asked.

'Sure,' said Retana. He took a fifty-peseta note out of his pocket-book and laid it, spread out flat, on the table.

Manuel picked it up and put it in his pocket.

'What about a cuadrilla?' he asked.

There's the boys that always work for me nights,' Retana said. 'They're all right.'

'How about picadors?' Manuel asked.

'They're not much,' Retana admitted.

'I've got to have one good pic,' Manuel said.

'Get him then,' Retana said. 'Go and get him.'

'Not out of this,' Manuel said. 'I'm not paying for any cuadrilla out of sixty duros.'

Retana said nothing but looked at Manuel across the big desk.

'You know I've got to have one good pic,' Manuel said.

Retana said nothing but looked at Manuel from a long way off.

'It isn't right,' Manuel said.

Retana was still considering him, leaning back in his chair, considering him from a long way away.

'There're the regular pics,' he offered.

'I know,' Manuel said. 'I know your regular pics.'

Retana did not smile. Manuel knew it was over.

'All I want is an even break,' Manuel said reasoningly. 'When I go out there I want to be able to call my shots on the bull. It only takes one good picador.'

He was talking to a man who was no longer listening.

'If you want something extra,' Retana said, 'go and get it. There will be a regular cuadrilla out there. Bring as many of your own pics as you want. The charlotada is over by 10.30.'

'All right,' Manuel said. 'If that's the way you feel about it.'

'That's the way,' Retana said.

'I'll see you tomorrow night,' Manuel said.

'I'll be out there,' Retana said.

Manuel picked up his suitcase and went out.

'Shut the door,' Retana called.

Manuel looked back. Retana was sitting forward looking at some papers. Manuel pulled the door tight until it clicked.

He went down the stairs and out of the door into the hot brightness of the street. It was very hot in the street and the light on the white buildings was sudden and hard on his eyes. He walked down the shady side of the street toward the Puerta del Sol. The shade felt solid and cool as running water. The heat came suddenly as he crossed the intersecting streets. Manuel saw no one he knew in all the people he passed.

Just before the Puerto del Sol he turned into a café.

It was quiet in the café. There were a few men sitting at tables against the wall. At one table four men played cards. Most of the men sat against the wall smoking, empty coffee-cups and liqueur-glasses before them on the tables. Manuel went through the long room to a small room in back. A man sat at a table in the corner asleep. Manuel sat down at one of the tables.

A waiter came in and stood beside Manuel's table.

'Have you seen Zurito?' Manuel asked him.

'He was in before lunch,' the waiter answered. 'He won't be back before five o'clock.'

'Bring me some coffee and milk and a shot of the ordinary,' Manuel said.

The waiter came back into the room carrying a tray with a big coffee-glass and a liqueur-glass on it. In his left hand he held a bottle of brandy. He swung these down to the table and a boy who had followed him poured coffee and milk into the glass from two shiny, spouted pots with long handles.

Manuel took off his cap and the waiter noticed his pigtail pinned forward on his head. He winked at the coffee-boy as he poured out the brandy into the little glass beside Manuel's coffee. The coffee-boy looked at Manuel's pale face curiously.

'You fighting here?' asked the waiter, corking up the bottle.

'Yes,' Manuel said. 'Tomorrow.'

The waiter stood there, holding the bottle on one hip.

'You in the Charlie Chaplins?' he asked.

The coffee-boy looked away, embarrassed.

'No. In the ordinary.'

'I thought they were going to have Chaves and Hernandez,' the waiter said.

'No. Me and another.'

'Who? Chaves or Hernandez?'

'Hernandez, I think.'

'What's the matter with Chaves?'

'He got hurt.'

'Where did you hear that?'

'Retana.'

'Hey, Looie,' the waiter called to the next room. 'Chaves got cogida.'

Manuel had taken the wrapper off the lumps of sugar and dropped them into his coffee. He stirred it and drank it down, sweet, hot, and warming in his empty stomach. He drank off the brandy.

'Give me another shot of that,' he said to the waiter.

The waiter uncorked the bottle and poured the glass full, slopping another drink into the saucer. Another waiter had come up in front of the table. The coffee-boy was gone.

'Is Chaves hurt bad?' the second waiter asked Manuel.

'I don't know,' Manuel said. 'Retana didn't say.'

'A hell of a lot he cares,' the tall waiter said. Manuel had not seen him before. He must have just come up.

'If you stand in with Retana in this town, you're a made man,' the tall waiter said. 'If you aren't in with him, you might just as well go out and shoot yourself.'

'You said it,' the other waiter who had come in said. 'You said it then.'

'You're right I said it,' said the tall waiter. 'I know what I'm talking about when I talk about that bird.'

'Look what he's done for Villalta,' the first waiter said.

'And that ain't all,' the tall waiter said. 'Look what he's done for Marcial Lalanda. Look what he's done for Nacional.'

'You said it, kid,' agreed the short waiter.

Manuel looked at them, standing talking in front of his table. He had drunk his second brandy. They had forgotten about him. They were not interested in him.

'Look at that bunch of camels,' the tall waiter went on. 'Did you ever see this Nacional II?'

'I seen him last Sunday, didn't I?' the original waiter said.

'He's a giraffe,' the short waiter said.

'What did I tell you?' the tall waiter said. 'Those are Retana's boys.'

'Say, give me another shot of that,' Manuel said. He had poured the brandy the waiter had slopped over in the saucer into his glass and drank it while they were talking.

The original waiter poured his glass full mechanically, and the three of them went out of the room talking.

In the far corner the man was still asleep, snoring slightly on the intaking breath, his head back against the wall.

Manuel drank his brandy. He felt sleepy himself. It was too hot to go out into the town. Besides there was nothing to do. He wanted to see Zurito. He would go to sleep while he waited. He kicked his suitcase under the table to be sure it was there. Perhaps it would be better to put it back under the seat, against the wall. He leaned down and shoved it under. Then he leaned forward on the table and went to sleep.

When he woke there was someone sitting across the table from him. It was a big man with a heavy brown face like an Indian. He had been sitting there some time. He had waved the waiter away and sat reading the paper and occasionally looking down at Manuel, asleep, his head on the table. He read the paper laboriously, forming the words with his lips as he read. When it tired him he looked at Manuel. He sat heavily in the chair, his black Cordoba hat tipped forward.

Manuel sat up and looked at him.

'Hello, Zurito,' he said.

'Hello, kid,' the big man said.

'I've been asleep,' Manuel rubbed his forehead with the back of his fist.

'I thought maybe you were.'

'How's everything?'

'Good. How is everything with you?'

'Not so good.'

They were both silent. Zurito, the picador, looked at Manuel's white face. Manuel looked down at the picador's enormous hands folding the paper to put away in his pocket.

'I got a favour to ask you, Manos,' Manuel said.

Manosduros was Zurito's nickname. He never heard it without thinking of his huge hands. He put them forward on the table self-consciously.

'Let's have a drink,' he said.

'Sure,' said Manuel.

The waiter came and went, and came again. He went out of the room looking back at the two men at the table.

'What's the matter, Manolo?' Zurito set down his glass.

'Would you pic two bulls for me tomorrow night?' Manuel asked, looking up at Zurito across the table.

'No,' said Zurito. 'I'm not pic-ing.'

Manuel looked down at his glass. He had expected that answer; now he had it. Well, he had it.

'I'm sorry, Manolo, but I'm not pic-ing.' Zurito looked at his hands.

'That's all right,' Manuel said.

'I'm too old,' Zurito said.

'I just asked you,' Manuel said.

'Is it the nocturnal tomorrow?'

'That's it. I figured if I had just one good pic, I could get away with it.'

'How much are you getting?'

'Three hundred pesetas.'

'I get more than that for pic-ing.'

'I know,' said Manuel. 'I didn't have any right to ask you.'

'What do you keep on doing it for?' Zurito asked. 'Why don't you cut off your coleta, Manolo?'

'I don't know,' Manuel said.

'You're pretty near as old as I am,' Zurito said.

'I don't know,' Manuel said. 'I got to do it. If I can fix it so that I get an even break, that's all I want. I got to stick with it, Manos.'

'No, you don't.'

'Yes, I do. I've tried keeping away from it.'

'I know how you feel. But it isn't right. You ought to get out and stay out.'

'I can't do it. Besides, I've been going good lately.'

Zurito looked at his face.

'You've been in the hospital.'

'But I was going great when I got hurt.'

Zurito said nothing. He tipped the cognac out of his saucer into his glass.

'The papers said they never saw a better faena,' Manuel said.

Zurito looked at him.

'You know when I get going I'm good,' Manuel said.

'You're too old,' the picador said.

'No,' said Manuel. 'You're ten years older than I am.'

'With me it's different.'

'I'm not too old,' Manuel said.

They sat silent, Manuel watching the picador's face.

'I was going great till I got hurt,' Manuel offered.

'You ought to have seen me, Manos,' Manuel said, reproachfully.

'I don't want to see you,' Zurito said. 'It makes me nervous.'

'You haven't seen me lately.'

'I've seen you plenty.'

Zurito looked at Manuel, avoiding his eyes.

'You ought to quit it, Manolo.'

'I can't,' Manuel said. 'I'm going good now, I tell you.'

Zurito leaned forward, his hands on the table.

'Listen. I'll pic for you and if you don't go big tomorrow night, you'll quit. See? Will you do that?'

'Sure.'

Zurito leaned back, relieved.

'You got to quit,' he said. 'No monkey business. You got to cut the coleta.'

'I won't have to quit,' Manuel said. 'You watch me. I've got the stuff.'

Zurito stood up. He felt tired from arguing.

'You got to quit,' he said. 'I'll cut your coleta myself.'

'No, you won't,' Manuel said. 'You won't have a chance.'

Zurito called the waiter.

'Come on,' said Zurito. 'Come on up to the house.'

Manuel reached under the seat for his suitcase. He was happy. He knew Zurito would pic for him. He was the best picador living. It was all simple now.

'Come on up to the house and we'll eat,' Zurito said.

* * *

Manuel stood in the patio de caballos waiting for the Charlie Chaplins to be over. Zurito stood beside him. Where they stood it was dark. The high door that led into the bull-ring was shut. Above them they heard a shout, then another shout of laughter. Then there was silence. Manuel liked the smell of the stables about the patio de caballos. It smelt good in the dark. There was another roar from the arena and then applause, prolonged applause, going on and on.

'You ever seen these fellows?' Zurito asked, big and looming beside Manuel in the dark.

'No,' Manuel said.

'They're pretty funny,' Zurito said. He smiled to himself in the dark.

The high, double, tight-fitting door into the bull-ring swung open and Manuel saw the ring in the hard light of the arc-lights, the plaza, dark all the way around, rising high; around the edge of the ring were running and bowing two men dressed like tramps, followed by a third in the uniform of an hotel bell-boy who stooped and picked up the hats and canes thrown down on to the sand and tossed them back up into the darkness.

The electric light went on in the patio.

'I'll climb on to one of these ponies while you collect the kids,' Zurito said.

Behind them came the jingle of the mules, coming out to go into the arena and be hitched on to the dead bull.

The members of the cuadrilla, who had been watching the burlesque from the runway between the barrera and the seats, came walking back and stood in a group talking, under the electric light in the patio. A good-looking lad in a silver-and-orange suit came up to Manuel and smiled.

'I'm Hernandez,' he said and put out his hand.

Manuel shook it.

'They're regular elephants we've got tonight,' the boy said cheerfully.

'They're big ones with horns,' Manuel agreed.

'You drew the worst lot,' the boy said.

'That's all right,' Manuel said. 'The bigger they are, the more meat for the poor.'

'Where did you get that one?' Hernandez grinned.

'That's an old one,' Manuel said. 'You line up your cuadrilla, so I can see what I've got.'

'You've got some good kids,' Hernandez said. He was very cheerful. He had been on twice before in nocturnals and was beginning to get a following in Madrid. He was happy the fight would start in a few minutes.

'Where are the pics?' Manuel asked.

'They're back in the corrals fighting about who gets the beautiful horses,' Hernandez grinned.

The mules came through the gate in a rush, the whips snapping, bells jangling and the young bull ploughing a furrow of sand.

They formed up for the paseo as soon as the bull had gone through.

Manuel and Hernandez stood in front. The youths of the cuadrillas were behind, their heavy capes furled over their arms. In back, the four picadors, mounted, holding their steel-tipped push-poles erect in the half-dark of the corral.

'It's a wonder Retana wouldn't give us enough light to see the horses by,' one picador said.

'He knows we'll be happier if we don't get too good a look at these skins,' another pic answered.

'This thing I'm on barely keeps me off the ground,' the first picador said.

'Well, they're horses.'

'Sure, they're horses.'

They talked, sitting their gaunt horses in the dark.

Zurito said nothing. He had the only steady horse of the lot. He had tried him, wheeling him in the corrals, and he responded to the bit and the spurs. He had taken the bandage off his right eye and cut the strings where they had tied his ears tight shut at the base. He was a good, solid horse, solid on his legs. That was all he needed. He intended to ride him all through the corrida. He had already, since he had mounted sitting in the half-dark in the big, quilted saddle, waiting for the paseo, pic-ed through the whole corrida in his mind. The other picadors went on talking on both sides of him. He did not hear them.

The two matadors stood together in front of their three peones, their capes furled over their left arms in the same fashion. Manuel was thinking about the three lads in back of him. They were all three Madrilenos, like Hernandez, boys about nineteen. One of them, a gipsy, serious, aloof, and dark-faced, he liked the look of. He turned.

'What's your name, kid?' he asked the gipsy.

'Fuentes,' the gipsy said.

'That's a good name,' Manuel said.

The gipsy smiled, showing his teeth.

'You take the bull and give him a little run when he comes out,' Manuel said.

'All right,' the gipsy said. His face was serious. He began to think about just what he would do.

'Here she goes,' Manuel said to Hernandez.

'All right. We'll go.'

Heads up, swinging with the music, their right arms swinging free, they stepped out, crossing the sanded arena under the arc-lights, the cuadrillas opening out behind, the picadors riding after, behind came the bull-ring servants and the jingling mules. The crowd applauded Hernandez as they marched across the arena. Arrogant, swinging, they looked straight ahead as they marched.

They bowed before the president, and the procession broke up into its component parts. The bull-fighters went over to the barrera and changed their heavy mantles for the light fighting capes. The mules went out. The picadors galloped jerkily around the ring, and two rode out the gate they had come in by. The servants swept the sand smooth.

Manuel drank a glass of water poured for him by one of Retana's deputies, who was acting as his manager and sword-handler. Hernandez came over from speaking with his own manager.

'You got a good hand, kid,' Manuel complimented him.

'They like me,' Hernandez said happily.

'How did the paseo go?' Manuel asked Retana's man.

'Like a wedding,' said the handler. 'Fine. You came out like Joselito and Belmonte.'

Zurito rode by, a bulky equestrian statue. He wheeled his horse and faced him toward the toril on the far side of the ring where the bull would come out. It was strange under the arc-light. He pic-ed in the hot afternoon sun for big money. He didn't like this arc-light business. He wished they would get started.

Manuel went up to him.

'Pic him, Manos,' he said. 'Cut him down to size for me.'

'I'll pic him, kid,' Zurito spat on the sand. 'I'll make him jump out of the ring.'

'Lean on him, Manos,' Manuel said.

'I'll lean on him,' Zurito said. 'What's holding it up?'

'He's coming now,' Manuel said.

Zurito sat there, his feet in the box-stirrups, his great legs in the buckskin-covered armour gripping the horse, the reins in his left hand, the long pic held in his right hand, his broad hat well down over his eyes to shade them from the lights, watching the distant door of the toril. His horse's ears quivered. Zurito patted him with his left hand.

The red door of the toril swung back and for a moment Zurito looked into the empty passageway far across the arena. Then the bull came out in a rush, skidding on his four legs as he came out under the lights, then charging in a gallop, moving softly in a fast gallop, silent except as he woofed through wide nostrils as he charged, glad to be free after the dark pen.

In the first row of seats, slightly bored, leaning forward to write on the cement wall in front of his knees, the substitute bull-fight critic of *El Heraldo* scribbled: 'Campagnero, Negro, 42, came out at 90 miles an hour with plenty of gas—'

Manuel, leaning against the barrera, watching the bull, waved his hand and the gipsy ran out, trailing his cape. The bull, in full gallop, pivoted and charged the cape, his head down, his tail rising. The gipsy moved in a zigzag, and as he passed, the bull caught sight of him and abandoned the cape to charge the man. The gip sprinted and vaulted the red fence of the barrera as the bull struck it with his horns. He tossed into it twice with his horns, banging into the wood blindly.

The critic of *El Heraldo* lit a cigarette and tossed the match at the bull, then wrote in his note-book, 'large and with enough horns to satisfy the cash customers, Campagnero showed a tendency to cut into the terrain of the bull-fighters.'

Manuel stepped out on the hard sand as the bull banged into the fence. Out of the corner of his eye he saw Zurito sitting the white horse close to the barrera, about a quarter of the way around the ring to the left. Manuel held the cape close in front of him, a fold in each hand, and shouted at the bull. 'Huh! Huh!' The bull turned, seemed to brace against the fence as he charged in a scramble, driving into the cape as Manuel side-stepped, pivoted on his heels with the charge of the bull, and swung the cape just ahead of the horns. At the end of the swing he was facing the bull again and held the cape in the same position close in front of his body, and pivoted again as the bull recharged. Each time, as he swung, the crowd shouted.

Four times he swung with the bull, lifting the cape so it billowed full, and each time bringing the bull around to charge again. Then, at the end of the fifth swing, he held the cape against his hip and pivoted, so the cape swung out like a ballet dancer's skirt and wound the bull around himself like a belt, to step clear, leaving the bull facing Zurito on the white horse, come up and planted firm, the horse facing the bull, its ears forward, its lips nervous, Zurito, his hat over his eyes, leaning forward, the long pole sticking out before and behind in a sharp angle under his right arm, held half way down, the triangular iron point facing the bull.

El Heraldo's second-string critic, drawing on his cigarette, his eyes on the bull, wrote: 'the veteran Manolo designed a series of acceptable veronicas, ending in a very Belmontistic recorte that earned applause from the regulars, and we entered the tercio of the cavalry.'

Zurito sat his horse, measuring the distance between the

bull and the end of the pic. As he looked, the bull gathered himself together and charged, his eyes on the horse's chest. As he lowered his head to hook, Zurito sunk the point of the pic in the swelling hump of muscle above the bull's shoulder, leaned all his weight on the shaft, and with his left hand pulled the white horse into the air, front hoofs pawing, and swung him to the right as he pushed the bull under and through so the horns passed safely under the horse's belly and the horse came down, quivering, the bull's tail brushing his chest as he charged the cape Hernandez offered him.

Hernandez ran sideways, taking the bull out and away with the cape, toward the other picador. He fixed him with a swing of the cape, squarely facing the horse and rider, and stepped back. As the bull saw the horse he charged. The picador's lance slid along his back, and as the shock of the charge lifted the horse, the picador was already half way out of the saddle, lifting his right leg clear as he missed with the lance and falling to the left side to keep the horse between him and the bull. The horse, lifted and gored, crashed over with the bull driving into him, the picador gave a shove with his boots against the horse and lay clear, waiting to be lifted and hauled away and put on his feet.

Manuel let the bull drive into the fallen horse; he was in no hurry, the picador was safe; besides, it did a picador like that good to worry. He'd stay on longer next time. Lousy pics! He looked across the sand at Zurito a little way out from the barrera, his horse rigid, waiting.

'Huh!' he called to the bull, 'Tomar!' holding the cape in both hands so it would catch his eye. The bull detached himself from the horse and charged the cape, and Manuel, running sideways and holding the cape spread wide, stopped, swung on his heels, and brought the bull sharply around facing Zurito.

'Campagnero accepted a pair of varas for the death of one rosinante, with Hernandez and Manolo at the quites,' *El Heraldo's* critic wrote. 'He pressed on the iron and clearly showed he was no horse-lover. The veteran Zurito resurrected some of his old stuff with the pike-pole, notably the suerte—'

'Olé! Olé!' the man sitting beside him shouted. The shout was lost in the roar of the crowd, and he slapped the critic on the back. The critic looked up to see Zurito, directly below him, leaning far out over his horse, the length of the pic rising in a sharp angle under his armpit, holding the pic almost by the point, bearing down with all his weight, holding the bull off, the bull pushing and driving to get at the horse, and Zurito, far out, on top of him, holding him, holding him, and slowly pivoting the horse against the pressure, so that at last he was clear. Zurito felt the moment when the horse was clear and the bull could come past, and relaxed the absolute steel lock of his resistance, and the triangular steel point of the pic ripped in the bull's hump of shoulder muscle as he tore loose to find Hernandez's cape before his muzzle. He charged blindly into the cape and the boy took him out into the open arena.

Zurito sat patting his horse and looking at the bull charging the cape that Hernandez swung for him out under the bright light while the crowd shouted.

'You see that one?' he said to Manuel.

'It was a wonder,' Manuel said.

'I got him that time,' Zurito said. 'Look at him now.'

At the conclusion of a closely turned pass of the cape the bull slid to his knees. He was up at once, but far out across the sand Manuel and Zurito saw the shine of the pumping flow of blood, smooth against the black of the bull's shoulder.

'I got him that time,' Zurito said.

'He's a good bull,' Manuel said.

'If they gave me another shot at him, I'd kill him,' Zurito said.

'They'll change the thirds on us,' Manuel said.

'Look at him now,' Zurito said.

'I got to go over there,' Manuel said, and started on a run for the other side of the ring, where the monos were leading a horse out by the bridle toward the bull, whacking him on the legs with rods and all, in a procession, trying to get him toward the bull, who stood, dropping his head, pawing, unable to make up his mind to charge.

Zurito, sitting his horse, walking him toward the scene, not missing any detail, scowled.

Finally the bull charged, the horse leaders ran for the barrera, the picador hit too far back, and the bull got under the horse, lifted him, threw him on to his back.

Zurito watched. The monos, in their red shirts, running out to drag the picador clear. The picador, now on his feet, swearing and flopping his arms. Manuel and Hernandez standing ready with their capes. And the bull, the great, black bull, with a horse on his back, hooves dangling, the bridle caught in the horns. Black bull with a horse on his back, staggering short-legged, then arching his neck and lifting, thrusting, charging to slide the horse off, horse sliding down. Then the bull into a lunging charge at the cape Manuel spread for him.

The bull was slower now, Manuel felt. He was bleeding badly. There was a sheen of blood all down his flank.

Manuel offered him the cape again. There he came, eyes open, ugly, watching the cape. Manuel stepped to the side and raised his arms, tightening the cape ahead of the bull for the veronica.

Now he was facing the bull. Yes, his head was going down a little. He was carrying it lower. That was Zurito.

Manuel flopped the cape; there he comes; he sidestepped and swung in another veronica. He's shooting awfully accurately, he thought. He's had enough fight, so he's watching now. He's hunting now. Got his eye on me. But I always give him the cape.

He shook the cape at the bull; there he comes; he sidestepped. Awful close that time. I don't want to work that close to him.

The edge of the cape was wet with blood where it had swept along the bull's back as he went by.

All right, here's the last one.

Manuel, facing the bull, having turned with him each charge, offered the cape with his two hands. The bull looked at him. Eyes watching, horns straight forward, the bull looked at him, watching.

'Huh!' Manuel said, 'Toro!' and leaning back, swung the

cape forward. Here he comes. He sidestepped, swung the cape in back of him, and pivoted, so the bull followed a swirl of cape and then was left with nothing, fixed by the pass, dominated by the cape. Manuel swung the cape under his muzzle with one hand, to show the bull was fixed, and walked away.

There was no applause.

Manuel walked across the sand toward the barrera, while Zurito rode out of the ring. The trumpet had blown to change the act to the planting of the banderillos while Manuel had been working with the bull. He had not consciously noticed it. The monos were spreading canvas over the two dead horses and sprinkling sawdust around them.

Manuel came up to the barrera for a drink of water. Retana's man handed him the heavy porous jug.

Fuentes, the tall gipsy, was standing holding a pair of banderillos, holding them together, slim, red sticks, fish-hook points out. He looked at Manuel.

'Go on out there,' Manuel said.

The gipsy trotted out. Manuel set down the jug and watched. He wiped his face with his handkerchief.

The critic of *El Heraldo* reached for the bottle of warm champagne that stood between his feet, took a drink, and finished his paragraph.

'—the aged Manolo rated no applause for a vulgar series of lances with the cape and we entered the third of the palings.'

Alone in the centre of the ring the bull stood, still fixed. Fuentes, tall, flat-backed, walking toward him arrogantly, his arms spread out, the two slim, red sticks, one in each hand, held by the fingers, points straight forward. Fuentes walked forward. Back of him and to one side was a peon with a cape. The bull looked at him and was no longer fixed.

His eyes watched Fuentes, now standing still. Now he leaned back, calling to him. Fuentes twitched the two banderillos and the light on the steel points caught the bull's eye.

His tail went up and he charged.

He came straight, his eyes on the man. Fuentes stood still, leaning back, the banderillos pointing forward. As the bull lowered his head to hook, Fuentes leaned backward, his arms came together and rose, his two hands touching, the banderillos two descending red lines, and leaning forward drove the points into the bull's shoulder, leaning far in over the bull's horns and pivoting on the two upright sticks, his legs tight together, his body curving to one side to let the bull pass.

'Olé!' from the crowd.

The bull was hooking wildly, jumping like a trout, all four feet off the ground. The red shaft of the banderillos tossed as he jumped.

Manuel, standing at the barrera, noticed that he looked always to the right.

'Tell him to drop the next pair on the right,' he said to the kid who started to run out to Fuentes with the new banderillos.

A heavy hand fell on his shoulder. It was Zurito.

'How do you feel, kid?' he asked.

Manuel was watching the bull.

Zurito leaned forward on the barrera, leaning the weight of his body on his arms. Manuel turned to him.

'You're going good,' Zurito said.

Manuel shook his head. He had nothing to do now until the next third. The gipsy was very good with the banderillos. The bull would come to him in the next third in good shape. He was a good bull. It had all been easy up to now. The final stuff with the sword was all he worried over. He did not really worry. He did not even think about it. But standing there he had a heavy sense of apprehension. He looked out at the bull, planning his faena, his work with the red cloth that was to reduce the bull, to make him manageable.

The gipsy was walking out toward the bull again, walking heel-and-toe, insultingly, like a ballroom dancer, the red shafts of the banderillos twitching with his walk. The bull watched him, not fixed now, hunting him, but waiting to get close enough so he could be sure of getting him, getting the horns into him.

As Fuentes walked forward the bull charged. Fuentes ran across the quarter of a circle as the bull charged and, as he passed running backward, stopped, swung forward, rose on his toes, arms straight out, and sunk the banderillos straight down into the tight of the big shoulder muscles as the bull missed him.

The crowd were wild about it.

'That kid won't stay in this night stuff long,' Retana's man said to Zurito.

'He's good,' Zurito said.

'Watch him now.'

They watched.

Fuentes was standing with his back against the barrera. Two of the cuadrilla were back of him, with their capes ready to flop over the fence to distract the bull.

The bull, with his tongue out, his barrel heaving, was watching the gipsy. He thought he had him now. Back against the red planks. Only a short charge away. The bull watched him.

The gipsy bent back, drew back his arms, the banderillos pointing at the bull. He called to the bull, stamped one foot. The bull was suspicious. He wanted the man. No more barbs in the shoulder.

Fuentes walked a little closer to the bull. Bent back. Called again. Somebody in the crowd shouted a warning.

'He's too damn close,' Zurito said.

'Watch him,' Retana's man said.

Leaning back, inciting the bull with the banderillos, Fuentes jumped, both feet off the ground. As he jumped the bull's tail rose and he charged. Fuentes came down on his toes, arms straight out, whole body arching forward, and drove the shafts straight down as he swung his body clear of the right horn.

The bull crashed into the barrera where the flopping capes had attracted his eye as he lost the man.

The gipsy came running along the barrera toward Manuel, taking the applause of the crowd. His vest was ripped where he had not quite cleared the point of the horn. He was happy about it, showing it to the spectators. He made the tour of the ring. Zurito saw him go by, smiling, pointing at his vest. He smiled.

Somebody else was planting the last pair of banderillos. Nobody was paying any attention.

Retana's man tucked a baton inside the red cloth of a muleta, folded the cloth over it, and handed it over the barrera to Manuel. He reached in the leather sword-case, took out a sword, and holding it by its leather scabbard, reached it over the fence to Manuel. Manuel pulled the blade out by the red hilt and the scabbard fell limp.

He looked at Zurito. The big man saw he was sweating.

'Now you get him, kid,' Zurito said.

Manuel nodded.

'He's in good shape,' Zurito said.

'Just like you want him,' Retana's man assured him.

Manuel nodded.

The trumpeter, up under the roof, blew for the final act, and Manuel walked across the arena toward where, up in the dark boxes, the president must be.

In the front row of seats the substitute bull-fight critic of *El Heraldo* took a long drink of the warm champagne. He had decided it was not worth while to write a running story and would write up the corrida back in the office. What the hell was it anyway? Only a nocturnal. If he missed anything he would get it out of the morning papers. He took another drink of the champagne. He had a date at Maxim's at twelve. Who were these bull-fighters anyway? Kids and bums. A bunch of bums. He put his pad of paper in his pocket and looked over toward Manuel, standing very much alone in the ring, gesturing with his hat in a salute toward a box he could not see high up in the dark plaza. Out in the ring the bull stood quiet, looking at nothing.

'I dedicate this bull to you, Mr. President, and to the public of Madrid, the most intelligent and generous of the world,' was what Manuel was saying. It was a formula. He said it all. It was a little long for nocturnal use.

He bowed at the dark, straightened, tossed his hat over his shoulder, and, carrying the muleta in his left hand and the sword in his right, walked out toward the bull.

Manuel walked toward the bull. The bull looked at him; his eyes were quick. Manuel noticed the way the banderillos hung down on his left shoulder and the steady sheen of blood from Zurito's pic-ing. He noticed the way the bull's feet were. As he walked forward, holding the muleta in his left hand and the sword in his right, he watched the bull's feet. The bull could not charge without gathering his feet together. Now he stood square on them, dully.

Manuel walked toward him, watching his feet. This was all right. He could do this. He must work to get the bull's head down, so he could go in past the horns and kill him. He did not think about the sword, not about killing the bull. He thought about one thing at a time. The coming things oppressed him, though. Walking forward, watching the bull's feet, he saw successively his eyes, his wet muzzle, and the wide, forward-pointing spread of his horns. The bull had light circles about his eyes. His eyes watched Manuel. He felt he was going to get this little one with the white face.

Standing still now and spreading the red cloth of the muleta with the sword, pricking the point into the cloth so that the sword, now held in his left hand, spread the red flannel like the jib of a boat, Manuel noticed the points of the bull's horns. One of them was splintered from banging against the barrera. The other was sharp as a porcupine quill. Manuel noticed while spreading the muleta that the white base of the horn was stained red. While he noticed these things he did not lose sight of the bull's feet. The bull watched Manuel steadily.

He's on the defensive now, Manuel thought. He's reserving himself. I've got to bring him out of that and get his head down. Always get his head down. Zurito had his head down once, but he's come back. He'll bleed when I start him going and that will bring it down.

Holding the muleta, with the sword in his left hand widening it in front of him, he called to the bull.

The bull looked at him.

He leaned back insultingly and shook the widespread flannel.

The bull saw the muleta. It was a bright scarlet under the arc-light. The bull's legs tightened.

Here he comes. Whoosh! Manuel turned as the bull came and raised the muleta so that it passed over the bull's horns and swept down his broad back from head to tail. The bull had gone clean up in the air with the charge. Manuel had not moved.

At the end of the pass the bull turned like a cat coming around a corner and faced Manuel.

He was on the offensive again. His heaviness was gone. Manuel noted the fresh blood shining down the black shoulder and dripping down the bull's leg. He drew the sword out of the muleta and held it in his right hand. The muleta held low down in his left hand, leaning toward the left, he called to the bull. The bull's legs tightened, his eyes on the muleta. Here he comes, Manuel thought. Yuh!

He swung with the charge, sweeping the muleta ahead of the bull, his feet firm, the sword following the curve, a point of light under the arcs.

The bull recharged as the pase natural finished and Manuel raised the muleta for a pase de pecho. Firmly planted, the bull came by his chest under the raised muleta. Manuel leaned his head back to avoid the clattering banderillo shafts. The hot, black bull body touched his chest as it passed.

Too damn close, Manuel thought. Zurito, leaning on the barrera, spoke rapidly to the gipsy, who trotted out toward Manuel with a cape. Zurito pulled his hat down low and looked out across the arena at Manuel.

Manuel was facing the bull again, the muleta held low and to the left. The bull's head was down as he watched the muleta.

'If it was Belmonte doing that stuff, they'd go crazy,' Retana's man said.

Zurito said nothing. He was watching Manuel out in the centre of the arena.

'Where did the boss dig this fellow up?' Retana's man asked.

'Out of the hospital,' Zurito said.

'That's where he's going damn quick,' Retana's man said.

Zurito turned on him.

'Knock on that,' he said, pointing to the barrera.

'I was just kidding, man,' Retana's man said.

'Knock on the wood.'

Retana's man leaned forward and knocked three times on the barrera.

'Watch the faena,' Zurito said.

Out in the centre of the ring, under the lights, Manuel was kneeling, facing the bull, and as he raised the muleta in both hands the bull charged, tail up.

Manuel swung his body clear and, as the bull recharged, brought around the muleta in a half-circle that pulled the bull off his knees.

'Why, that one's a great bull-fighter,' Retana's man said.

'No, he's not,' said Zurito.

Manuel stood up and, the muleta in his left hand, the sword in his right, acknowledged the applause from the dark plaza.

The bull had humped himself up from his knees and stood waiting, his head hung low.

Zurito spoke to two of the other lads of the cuadrilla and they ran out to stand back of Manuel with their capes. There were four men back of him now. Hernandez had followed him since he first came out with the muleta. Fuentes stood watching, his cape held against his body, tall in repose, watching, lazy-eyed. Now the two came up. Hernandez motioned them to stand one at each side. Manuel stood alone, facing the bull.

Manuel waved back the men with the capes. Stepping back cautiously, they saw his face was white and sweating. Didn't they know enough to keep back? Did they want to catch the bull's eye with the capes after he was fixed and ready? He had enough to worry about without that kind of thing.

The bull was standing, his four feet square, looking at the muleta. Manuel furled the muleta in his left hand. The bull's eyes watched it. His body was heavy on his feet. He carried his head low, but not too low.

Manuel lifted the muleta at him. The bull did not move. Only his eyes watched.

He's all lead, Manuel thought. He's all square. He's framed right. He'll take it.

He thought in bull-fight terms. Sometimes he had a thought and the particular piece of slang would not come into his mind and he could not realize the thought. His instincts and his knowledge worked automatically, and his brain worked slowly and in words. He knew all about bulls. He did not have to think about them. He just did the right thing. His eyes noted things and his body performed the necessary measures without thought. If he thought about it, he would be gone.

Now, facing the bull, he was conscious of many things at the same time. There were the horns, the one splintered, the other smoothly sharp, the need to profile himself toward the left horn, lance himself short and straight, lower the muleta, so the bull would follow it, and, going in over the horns, put the sword all the way into a little spot about as big as a five-peseta piece straight in back of the neck, between the sharp pitch of the bull's shoulders. He must do all this and must then come out from between the horns. He was conscious he must do all this, but his only thought was in words: 'Corto y derecho.'

'Corto y derecho,' he thought, furling the muleta. Short and straight. Corto y derecho, he drew the sword out of the muleta, profiled on the splintered left horn, dropped the muleta across his body, so his right hand with the sword on the level with his eye made the sign of the cross, and, rising on his toes, sighted along the dipping blade of the sword at the spot high up between the bull's shoulders.

Corto y derecho he launched himself on the bull.

There was a shock, and he felt himself go up in the air. He pushed on the sword as he went up and over, and it flew out of his hand. He hit the ground and the bull was on him. Manuel, lying on the ground, kicked at the bull's muzzle with his slippered feet. Kicking, kicking, the bull after him, missing him in his excitement, bumping him with his head, driving the horns into the sand. Kicking like a man keeping a ball in the air, Manuel kept the bull from getting a clean thrust at him.

Manuel felt the wind on his back from the capes flopping at the bull, and then the bull was gone, gone over him in a rush. Dark, as his belly went over. Not even stepped on.

Manuel stood up and picked up the muleta. Fuentes handed him the sword. It was bent where it had struck the shoulder-blade. Manuel straightened it on his knee and ran toward the bull, standing now beside one of the dead horses. As he ran, his jacket flopped where it had been ripped under his armpit.

'Get him out of there,' Manuel shouted to the gipsy. The bull had smelled the blood of the dead horse and ripped into the canvas cover with his horns. He charged Fuentes's cape, with the canvas hanging from his splintered horn, and the crowd laughed. Out in the ring, he tossed his head to rid himself of the canvas. Hernandez, running up from behind him, grabbed the end of the canvas and neatly lifted it off the horn.

The bull followed it in a half-charge and stopped still. He was on the defensive again. Manuel was walking toward him with the sword and muleta. Manuel swung the muleta before him. The bull would not charge.

Manuel profiled toward the bull, sighting along the dipping blade of the sword. The bull was motionless, seemingly dead on his feet, incapable of another charge.

Manuel rose to his toes, sighting along the steel, and charged.

Again there was the shock and he felt himself being borne back in a rush, to strike hard on the sand. There was no chance of kicking this time. The bull was on top of him. Manuel lay as though dead, his head on his arms, and the bull bumped him. Bumped his back, bumped his face in the sand. He felt the horn go into the sand between his folded arms. The bull hit him in the small of the back. His face drove into the sand. The horn drove through one of his sleeves and the bull ripped it off. Manuel was tossed clear and the bull followed the capes.

Manuel got up, found the sword and muleta, tried the point of the sword with his thumb, and then ran toward the barrera for a new sword.

Retana's man handed him the sword over the edge of the barrera.

'Wipe off your face,' he said.

Manuel, running again toward the bull, wiped his bloody face with his handkerchief. He had not seen Zurito. Where was Zurito?

The cuadrilla had stepped away from the bull and waited with their capes. The bull stood, heavy and dull again after the action.

Manuel walked toward him with the muleta. He stopped and shook it. The bull did not respond. He passed it right and left, left and right before the bull's muzzle. The bull's eyes watched it and turned with the swing, but he would not charge. He was waiting for Manuel.

Manuel was worried. There was nothing to do but go in. Corto y derecho. He profiled close to the bull, crossed the muleta in front of his body and charged. As he pushed in the sword, he jerked his body to the left to clear the horn. The bull passed him and the sword shot up in the air, twinkling under the arc-lights, to fall red-hilted on the sand.

Manuel ran over and picked it up. It was bent and he straightened it over his knee.

As he came running toward the bull, fixed again now, he passed Hernandez standing with his cape.

'He's all bone,' the boy said encouragingly.

Manuel nodded, wiping his face. He put the bloody handkerchief in his pocket.

There was the bull. He was close to the barrera now. Damn him. Maybe he was all bone. Maybe there was not any place for the sword to go in. The hell there wasn't! He'd show them.

He tried a pass with the muleta and the bull did not move. Manuel chopped the muleta back and forth in front of the bull. Nothing doing.

He furled the muleta, drew the sword out, profiled and drove it on the bull. He felt the sword buckle as he shoved it in, leaning his weight on it, and then it shot high in the air, end-over-ending into the crowd. Manuel had jerked clear as the sword jumped.

The first cushions thrown down out of the dark missed him. Then one hit him in the face, his bloody face looking toward the crowd. They were coming down fast. Spotting the sand. Somebody threw an empty champagne bottle from close range. It hit Manuel on the foot. He stood there watching the dark, where the things were coming from. Then something whisked through the air and struck by him. Manuel leaned over and picked it up. It was his sword. He straightened it over his knee and gestured with it to the crowd.

'Thank you,' he said. 'Thank you!'

Oh, the dirty bastards! Dirty bastards! Oh, the lousy, dirty bastards! He kicked into a cushion as he ran.

There was the bull. The same as ever. All right, you dirty, lousy bastard!

Manuel passed the muleta in front of the bull's black muzzle.

Nothing doing.

You won't! All right. He stepped close and jammed the sharp peak of the muleta into the bull's damp muzzle.

The bull was on him as he jumped back and as he tripped on a cushion he felt the horn go into him, into his side. He grabbed the horn with his two hands and rode backward, holding tight on to the place. The bull tossed him and he was clear. He lay still. It was all right. The bull was gone.

He got up coughing and feeling broken and gone. The dirty bastards!

'Give me the sword,' he shouted. 'Give me the stuff.'

Fuentes came up with the muleta and the sword.

Hernandez put his arm around him.

'Go on to the infirmary, man,' he said. 'Don't be a damn fool.'

'Get away from me,' Manuel said. 'Get to hell away from me.'

He twisted free. Hernandez shrugged his shoulders. Manuel ran toward the bull.

There was the bull standing, heavy, firmly planted.

All right, you bastard! Manuel drew the sword out of the muleta, sighted with the same movement, and flung himself on to the bull. He felt the sword go in all the way. Right up to the guard. Four fingers and his thumb into the bull. The blood was hot on his knuckles, and he was on top of the bull.

The bull lurched with him as he lay on, and seemed to sink; then he was standing clear. He looked at the bull going down slowly over on his side, then suddenly four feet in the air.

Then he gestured at the crowd, his hand warm from the bull blood.

All right, you bastards! He wanted to say something, but he started to cough. It was hot and choking. He looked down for the muleta. He must go over and salute the president. President hell! He was sitting down looking at something. It was the bull. His four feet up. Thick tongue out. Things crawling around on his belly and under his legs. Crawling where the hair was thin. Dead bull. To hell with the bull! To hell with them all! He started to get to his feet and commenced to cough. He sat down again, coughing. Somebody came and pushed him up.

They carried him across the ring to the infirmary, running with him across the sand, standing blocked at the gate as the mules came in, then around under the dark passageway, men grunting as they took him up the stairway, and then laid him down.

The doctor and two men in white were waiting for him. They laid him out on the table. They were cutting away his shirt. Manuel felt tired. His whole chest felt scalding inside. He started to cough and they held something to his mouth. Everybody was very busy.

There was an electric light in his eyes. He shut his eyes.

He heard someone coming very heavily up the stairs. Then he did not hear it. Then he heard a noise far off. That was the crowd. Well, somebody would have to kill his other bull. They had cut away all his shirt. The doctor smiled at him. There was Retana.

'Hello, Retana!' Manuel said. He could not hear his voice.

Retana smiled at him and said something. Manuel could not hear it.

Zurito stood beside the table, bending over where the doctor was working. He was in his picador clothes, without his hat.

Zurito said something to him. Manuel could not hear it.

Zurito was speaking to Retana. One of the men in white smiled and handed Retana a pair of scissors. Retana gave them to Zurito. Zurito said something to Manuel. He could not hear it.

To hell with this operating-table. He'd been on plenty of operating-tables before. He was not going to die. There would be a priest if he was going to die.

Zurito was saying something to him. Holding up the scissors.

That was it. They were going to cut off his coleta. They were going to cut off his pigtail.

Manuel sat up on the operating-table. The doctor stepped back, angry. Someone grabbed him and held him.

'You couldn't do a thing like that, Manos,' he said.

He heard suddenly, clearly, Zurito's voice.

'That's all right,' Zurito said. 'I won't do it. I was joking.'

'I was going good,' Manuel said. 'I didn't have any luck. That was all.'

Manuel lay back. They had put something over his face. It was all familiar. He inhaled deeply. He felt very tired. He was very, very tired. They took the thing away from his face.

'I was going good,' Manuel said weakly, 'I was going great.'

Retana looked at Zurito and started for the door.

'I'll stay here with him,' Zurito said.

Retana shrugged his shoulders.

Manuel opened his eyes and looked at Zurito.

'Wasn't I going good, Manos?' he asked, for confirmation.

'Sure,' said Zurito. 'You were going great.'

The doctor's assistant put the cone over Manuel's face and he inhaled deeply. Zurito stood awkwardly, watching.

ANNOTATIONS

Abbreviations

Am. = American
coll. = colloquial (Umgangssprache)
sl. = slang
Span. = Spanish

ABOUT THE AUTHOR

5 **5** *to graduate* to pass final examination at high school or college (Schule oder Universität erfolgreich abschließen, hier etwa: das Abitur machen)
high school secondary school (Gymnasium)

6 *Kansas City* on the mouth of the Kansas River into the Missouri, about 600.000 inhabitants

7 *prize-fighter* boxer (for money)
gunman (Am.) armed robber.

6 **7** *Stein*, Gertrude (1874—1946), author of 'Three Lives' (1908) and 'Making of Americans' (1925), has had a great influence on contemporary literature. She is said to have been 'the fountainhead for all the young Americans who were writing in Paris after the first World War'. Hemingway and artists of the 'advanced' group (Picasso, Matisse, Braque) were among her friends
Turgenev, Ivan Sergievich (1818—83), one of the greatest of Russian novelists. His most important novels are 'A Nest of Gentlefolk' (1859), 'Fathers and Sons' (1862), 'Smoke' (1867), and 'Virgin Soil' (1876)
Anderson, Sherwood (1876—1941), one of the foremost of the modern American school of novelists. His chief works are 'Marching Men' (1917), 'Dark Laughter' (1925), 'Hello Towns' (1929). He writes in a mood of sombre realism

8 *Twain*, Mark (1835—1910). His real name was Samuel Langhorne Clemens. He adopted the pseudonym from the leadsman's call which had become familiar to him on the Mississippi. His best-known works are 'Tom Sawyer' (1876) and 'Huckleberry Finn' (1884), a masterpiece of humorous fiction and an accurate picture of the old rough civilization of the Mississippi. Of his short stories 'The Celebrated Jumping Frog' (1864) first brought him into prominence as a writer. Whereas the short stories he wrote when he was young are based on tall tales of the West, the ones he wrote when he was old, e. g. 'The Canvasser's Tale', are based on his

disillusionment with the human race
to evolve to develop

11 *understatement* to express something negatively, by less than it actually deserves (Unterbewertung).

THE KILLERS

7 3 *counter* table on which money is counted out and across which goods are delivered (Ladentisch, Theke)
4 *What's yours?* (coll.) What would you like to have?
6 *Al* diminutive of *Alec(k)* ['ælik], *Alexander* [ælig'zɑːndə]
9 *the light came on* appeared, became visible
11 *menu* ['menjuː] bill of fare, list of dishes
Nick diminutive of *Nicholas*
14 *tenderloin* (Am.) (Lendenstück)
apple sauce (Apfelmus)
15 *mashed potatoes* (Kartoffelbrei, ~püree)
17 *the hell* one of the many swear-words you will come across in the text (etwa: zum Teufel)
28 *sandwich* two slices of bread with meat or something else between
31 *croquette* seasoned and fried ball of rice, potato, or meat (in this case: chicken-meat), etc.

8 5 *derby hat* stiff felt hat usually with a dome-shaped crown (steifer Hut; Melone)
7 *muffler* scarf (Halstuch)
13 *got anything?* (coll.) have you got anything?
14 *silver beer* cider (Apfelsaft, ~most)
bevo ['bevou] a lemonade
ginger-ale ginger-flavoured drink (*ginger* = Ingwer)
15 *you got* (coll.) have you got
17 *hot* in this context can have various meanings: 1. very warm — because they are asking for a drink; 2. acrid, sharp, biting — because of the drinks offered (*ginger* is hot, for instance). In connection with *silver beer* and *bevo* the word *hot* would have an ironical meaning; 3. still warm, fresh, new — because they had never heard of the town yet; 4. violent, vehement, impetuous, exciting — because of the things that are going to happen
19 *Summit* town in the State of New Jersey, south of New York, known as a railway junction and health resort. Otherwise the word *summit* means the highest point, the top of a mountain for instance. Why, do you think, did Hemingway choose this particular town for his story?

20 *ever hear of it?* (coll.) have you ever heard of it?
22 *nights* by night, every night, every evening
25/26 Think about the different meanings of *that's right*
27 *sure* (Am.) surely, certainly, of course
28 *bright* (ironically used) quick-witted, clever
34 *ain't* (sl.) isn't
37 *platter* flat dish or plate, often of wood.

9 1 *side-dish* extra dish
wicket small door, gate or opening
15 *see?* (coll.) do you see (understand)?
18 *a good one* a good joke.

10 1 *silly* adjectives instead of adverbs (in a silly way) are very common in colloquial American
2 *to argue* ['ɑ:gju:] to discuss
22 *back of* on the other (opposite) side of, behind
to make over to change, turn
23 *saloon* (Am.) drinking-bar
39 *to prop open* to hold up.

11 1 *catsup or ketchup* sauce made from juice of tomatoes, etc.
11 *don't he?* (sl.) doesn't he?
16 *ever go?* (coll.) do you ever go?
movies cinema, films
17 *once in a while* very rarely
22 *he never even seen us* (coll.) he has never even seen us
27 *to oblige* to please, do a favour to
29 *goddam* swear-word like *damn, the hell,* etc.
34 *to tie up* to bind together
38 *kosher* ['kouʃə] fulfilling requirements of Jewish law (koscher, rein). The fact that Max uses the word *kosher* may suggest that he is a Jew. Moreover he is rather talkative.

12 2 *to keep after it* (coll.) to insist on having something to eat
3 *get* = understand
4 *what you going to do?* (coll.) what are you going to do?
9 *street-car* (Am.) tram
motorman driver
16 *regular* real
18 *to blow somebody's head off* to shoot somebody
28 *ledge* upper edge or bar of the wicket
34 *you'd make some girl a nice wife* ironical allusion to the fact that wives in America and England are often spoiled by their husbands, who even wash up and cook food for the family.

13 14 *to be through with* to have done, be finished with
15 *sloppy* unsystematic, not thorough, unsafe
16 *to keep amused* to have a little fun
20 *bulge* irregular swelling
24 *to play the races* to bet on horses
26 *out the door* (Am.) out of the door
27 *arc-light* light made by arc, a luminous bridge formed by two separate carbon poles (Bogenlampe)
29 *vaudeville* ['voudəvil] variety entertainment
36 *to swagger* to walk like a superior, behave in a domineering way, talk boastfully; *to swagger something off* (etwas prahlerisch, angeberisch abtun, etwas nicht wahr haben wollen).

14 4 *they all gone?* (coll.) are they all gone?
5 *yeah* [jiæ] (Am.) yes
6 *any* anything
8 *go see* go and see
12 *way out of it* far, quite out of it. The way Sam reacts does not necessarily prove that he is a coward, that he lacks personal courage. In everything he does or says he has to consider that he is living among white people as a coloured man, he is, therefore, doubly handicapped, he runs a double risk
14 *it ain't going to get you anywhere* (coll.) it won't be any good, will be of no use
20 *rooming-house* boarding-house.

15 1 *prize-fighter* boxer for money, professional boxer
11 *you come in* (sl.) you came in
31 *I been* (coll.) I have been
38 *to fix up* (Am.) to arrange, organize.

16 1 *I got in wrong* I am done for, it's all up with me (ich liege falsch)
11 *I guess* (Am.) I suppose
12 *fall* (Am.) autumn
16 *ring* boxing-ring or ring (group, clique) of gangsters
20 *just as* ever so.

17 8 *to double-cross* (Am.) to cheat, betray each of two parties (betrügen, verpetzen, verraten)
10 *this town* compare note on the word *hot* on the second page of the story. Nick is most upset of all. It was probably the first time in his young life that he drifted into the sphere of evil. The personal courage he shows by warning Ole Andreson, though largely due to his inexperience with the world of evil, is about the only thing

that mitigates the sinister impression of utter hopelessness, futility, and despair the story leaves on the reader.

A DAY'S WAIT

18 *to ache* [eik] to hurt
28 *fever* ['fi:və] temperature
35 *capsule* ['kæpsju:l] gelatine envelope enclosing pill (Kapsel)
36 *purgative* ['pə:gətiv] medicine ['medsin] serving to purify (Abführmittel).

18 **1** *acid* sour; *to overcome an acid condition* to get rid of increased acidity in stomach (Überschuß an Magensäure beseitigen)
5 *'flu* short for *influénza* (Grippe)
6 *pneumonia* [nju:'mounjə] inflammation of the lungs
12 *detached* far away
14 *Howard Pyle* American author
22 *strangely* the adverb *strangely* indicates that the way of looking was strange, not that he was (looked) strange. In the latter case you have to use the adjective: to look happy, tired, etc.
27 *it* the key-word (Schlüsselwort) in this story. Father and son understand quite different things by *it*
31 *light-headed* delirious, confused (wirr im Kopf)
35 *sleet* hail or snow mixed with rain (Schloßen, Graupeln)
36 *brush* (Am.) brushwood (Unterholz)
37 *to varnish* to apply glazy or glossy surface to (lackieren, polieren, überziehen)
38 *setter* long-haired dog trained to stand rigid in scouting game (Vorstehhund)
39 *creek* tributary river, brook (Bach).

19 **2** *having it slide* letting it glide
4 *to flush* to cause to fly up (aufjagen)
covey ['kʌvi] set (Kette)
quail kind of partidge (Wachtel)
6 *to light, lit, lit* to alight (sich niederlassen)
8 *mound* artificial hill, heap
10 *to be poised* to balance
35 *to hold tight on to oneself about something* to keep something to oneself, to keep something secret
38 *to do any good* to help in some way.

20 **22** *like how* (Am.) how
25 *to relax* to grow less tense (nachlassen)

26 *hold* tension, power (Spannung, Verkrampfung)
27 *slack* relaxed, little.

THE BATTLER

29 *battler* fighter
30 *track* line of railway
31 *caboose* (Am.) brakevan (Bremserwagen)
33 *tamarack* American larch (Lärche)
34 *to feel of* to feel
pants (Am.) trousers
35 *to bark* to strip bark (Rinde) from tree, to tear off
36 *cinders* ashes.

21 3 *to squat* to sit with knees drawn up
5 *crut* (sl.) filthy fellow (Schweinehund)
7 *kid* (sl.) child
8 *to fall for* (sl.) to yield to the charms of (hereinfallen auf)
10 *to suck* (sl.) to take in, deceive, cheat
12 *to light*, *lit*, *lit* to land
17 *gotten* (Am.) got
22 *embankment* raised stone structure (Damm)
23 *to ballast* to fill with ballast (material used for forming railroad bed)
24 *tie* (Am.) rail sleeper (Bahnschwelle)
25 *causeway* ['kɔːzwei] or *causey* ['kɔːzei] raised road across low or wet place (Dammweg)
28 *freight* [freit] *train* goods train
29 *yard* space near station where rolling-stock is kept, trains are made up, etc. (Rangierbahnhof)
junction station where railway-lines meet
33 *so* so that
35 *to hike* to walk
back of behind
39 *slit* opening
spike large stout nail especially as used for railways

22 11 *beechnut burr* cover, husk of beechnut (Bucheckerhülse)
14 *he looked to be alone* he seemed to be alone
22 *shiner* something that shines, that you can see; *to get the shiner* (coll.) to be hurt (zugerichtet werden)
23 *to bust* (sl.) to throw (out)
24 *through freight* goods train that does not stop at smaller stations
26 *bastard* filthy fellow

27 *'bout* about
36 *a tough one* (Am.) brawler, rowdy (Raufbold, ein ‚Gefährlicher').

23 3 *misshapen* ill-shaped, deformed
sunken sunk
6 *to mutilate* to deform
putty ['pʌti] used for fixing panes of glass, filling up holes in wood work, etc. (Kitt)
8 *pan* (sl.) face
15 *ever see?* did you ever see?
16 *sick* unwell
19 *you bet* (coll.) of course, certainly
24 *Ad* diminutive of *Adolph*
31 *you ever been?* (coll.) have you ever been?
37 *honest to God?* (coll.) really?

24 8 *to bulge* to swell outwards
18 *throb* palpitation, pulsation, beating
24 *she* the heart, here regarded as a machine, so that the feminine gender is grammatically justified, the same as with ships, aeroplanes, etc. Note, however, that on page 25 the word heart is dealt with as a neuter. Try to explain the difference
27 *Bugs* nickname; a *bug* is a flat, ill-smelling, blood-sucking insect infesting beds (Wanze)
32 *pal* [pæl] (coll.) friend, comrade
33 *you say?* (coll.) did you say?

25 1 *he's got a lot coming to him* there is still a lot in store for him (ihm steht noch allerlei bevor)
5 *right away* now, in a minute
7 *hungry as hell* (sl.) as hungry as a wolf
8 *hear that?* (coll.) did you hear that?
12 *skillet* pan with long handle and usually legs used in cooking (Bratpfanne)
15 *to baste* to moisten with fat to prevent burning (mit Fett überziehen)
23 *to hang on to* not to let go, to keep
30 *you bet!* here meaning more than 'of course, surely, certainly'. The English equivalent is something like 'gladly', 'with pleasure', 'thank you'
38 *to watch out* to mind (aufpassen)
39 *remainder* what is left.

26 4 *right* really, very
20 *how do you get that way?* why did you suddenly change like that?
22 *snotty* (vulgar) dirty, filthy, mean

27 *a hot sketch* (sl.) mean fellow (Schweinehund)
to butt to run
31 *to act snotty* (sl.) to behave badly
32 *to talk snotty* (sl.) to talk insultingly
33 *Where the hell do you think you get off?* (vulgar) What do you think is going to happen to you?
36 *can* (sl.) head

27 3 *to get out of it* to get off with it
4 *lead at me* start, begin
5 *cut it out* (coll.) stop talking like that, leave it
9 *to set oneself* to start
10 *the base of the skull* (die Schädelbasis)
11 *blackjack* short stick with heavily loaded end (Stock mit Bleiknopf, Totschläger)
24 *flexible* that will bend without breaking, pliable (biegsam)
limber flexible, pliable (geschmeidig)
27 *whalebone* bone of a whale (Fischbein)
29 *to take care of* here: to defend
30 *no more* any more. The double negation is typical of the coloured man's way of speaking. A little later he says *nothing* instead of 'anything' and *none* instead of 'in any way' or 'at all'. Abbreviated grammatical forms such as *I seen* in the same line for 'I've seen' are equally characteristic of this mode of speaking, similarly *they was* for 'they were' on page 28.

28 6 *to sip* to drink in repeated tiny mouthfuls or by spoonfuls (schlürfen)
simple uncomplicated, primitive
12 *I remember about it* (coll.) I remember it vaguely
14 *didn't* who did not, see note to page 27, line 30
16 *come* came, see same note
23 *awful* see note to page 10
25 *to bust* to batter, beat up
28 *jail* prison
30 *cuttin'* cutting
32 *to look somebody up* to go to see somebody
35 *larceny* stealing, theft (Diebstahl).

29 2 *though* it is true, however
5 *mighty* (coll.) very
11 *I wish you'd sort of pull out* (coll.) I wish you would disappear, go away, if you don't mind
12 *to not be* instead of grammatically correct 'not to be'
back again when he has regained consciousness

13 *to thump* to beat heavily
19 *to hit a town* to come to a town
33 *to soothe* to calm
just you drink just drink. The additional *you* sounds affectionate
36 *mounting grade* rising slope.

AFTER THE STORM

30 2 *punch* drink of wine or spirits mixed with hot water, sugar, lemon, etc.
4 *to choke* to squeeze throat
5 *like he was* as if he were
6 *to cut him loose* to cut him so that he let loose
9 *muscle* ['mʌsl]
10 *he let go of me* he let loose
13 *you want* did you want
14 *I'd* I should
15 *bad* instead of 'badly'
22 *broke* broken
24 *skiff* light rowing-boat
25 *inside of* inside, in
Mango Key all the names mentioned in this story are connected with the *Keys* off the southern tip of Florida and Havana (Cuba), on the other side of Florida Strait. The exact location of places (such as *Mango Key*) is not given in these notes as they are of minor or almost no importance for understanding the story. *Key*, from Spanish *cayo* meaning 'shoal' (Sandbank), is here used in the original meaning of the word
26 *to bail* (or *to bale*) to throw water out of boat with pails or buckets
28 *plenty rough* very rough
I took it down along I put up with it all the time (ich fand mich wohl oder übel die ganze Zeit damit ab); *to take something down* to swallow something with difficulty
32 *lye* [lai] alkalized water (Lauge).

31 5 *I seen* I saw
spar stout pole such as used for mast
7 *schooner* ['sku:nə] sailing-ship with two or more masts (Schoner)
10 *to have the start on* to be the first of
14 *quicksands* loose wet sand easily swallowing up ships, animals, etc. (Triebsand, Flugsand)
16 *Rebecca Light* a lighthouse
20 *I got over close* I got near
23 *I come* I came

30 *laying* instead of the grammatically correct form 'lying'
33 *to scull* to propel boat with sculls or oars
35 *water glass* tube with glass bottom enabling objects under water to be observed
like we use sponging as we use them for gathering sponges
36 *so I could* so that I sould.

32 6 *rounded over* turned over by waves
8 *tackle* ['tɑ:kl] windlass with its ropes and hooks (Takelage)
13 *grains pole* forked fish-spear or harpoon
15 *wrench* instrument made to grip and turn nuts, bolts, etc. (Schraubenschlüssel)
to lash to fasten with cord
21 *shaky* nervous, excited
26 *I dove* 'to dive' is properly a weak verb, here the author makes it a strong one
33 *dinghy* ['diŋgi] small ship's boat.

33 17 *to quit* to stop
23 *shark* large voracious sea-fish (Hai)
25 *grapple* instrument for holding (Enterhaken)
32 *was floated* instead of 'had floated'
34 *plenty glad* (coll.) thoroughly, very glad
39 *it didn't do any good* it was of no use, without any result.

34 11 *to tow* [tou] to pull along in water by rope or chain (ins Schlepptau nehmen)
14 *it came on to blow* it started blowing
18 *bond* binding engagement (Bürgschaft, Kaution)
20 *Greek* here meaning 'cunning person', 'pirate'
22 *dynamite* ['dainəmait]
24 *nickel* five-cent piece
26 *Havana* capital of Cuba, Hemingway's temporary home
27 *to chance* to take the chance or risk
28 *to go with it* to go with the hurricane
30 *Tortugas* little group of coral-islands in the Gulf of Mexico belonging to Florida, west of the barrier-reef of the Key Islands. There are lighthouses on Tortugas
35 *ballast tanks* (Schotten)
lay instead of the correct verb 'lie'
37 *on her beam ends* on her side.

35 7 *jewfish* large marine fish. The species ['spi:ʃi:z] of the West Indies and the west coast of Mexico is 'Promicrops itaiara'
18 *to scud* to run before the wind (lenzen; vor schwerem Sturm mit

stark gerefften Segeln laufen)

35 *mate* second officer on ship (Maat, Steuermann)

39 *took* instead of correct 'taken'.

36 2 *they picked her clean* they took everything out of her

there was instead of correct 'there were'

THE DOCTOR AND THE DOCTOR'S WIFE

6 *Dick* diminutive of *Richard*

7 *Eddy* diminutive of *Edward*

8 *Billy* diminutive of *William*

10 *cross-cut saw* saw used or made for cross-cutting, cutting across the grain of wood (Schrotsäge)

to flop to sway heavily (schwingen, schwappen)

12 *cant-hook* a wooden lever with a movable iron hook near the end used for canting or turning over logs, etc. (Kanthaken)

16 *boom* floating barrier of logs or timber (Holzfloß)

17 *to tow* [tou] to pull along in water by rope or chain

18 *were done* is subjunctive; why?

26 *to waterlog* to saturate or fill with water as barely to float (sich voll Wasser saugen)

27 *to assume* to suppose

30 *wedge* piece of wood or metal used for splitting wood (Keil)

cord wood measure of cut wood, usually 128 cubic feet (Klafterholz)

chunk thick lump cut off (dickes Stück)

34 *crotch* angle formed by the parting of two branches or where the trunk of the tree divides (Gabelung)

35 *dock* leather case

half-breed offspring of Indian and white parents (Mischling).

37 2 *plug* piece of pressed tobacco cut off for chewing (Prieme)

3 *Ojibway* [ou'dʒibwei] largest Indian tribe north of Mexico. According to their tradition, they emigrated from the St. Lawrence region to that about the Great Lakes, especially Lake Superior, where most of them still live on several reservations in the United States and Canada. Chiefly hunters, mat weavers, and canoe builders. Their mythology inspired Longfellow to write his famous poem 'Hiawatha'

14 *to rock* to move gently to and fro (rütteln)

19 *who it belongs to* instead of *whom* ... 'who' instead of 'whom' in relative and interrogative sentences is commonly accepted nowadays

23 *scaler* man who measures and determines the volume of logs and stamps each log with the owner's mark

28 *buffy* easily offended

33 *half-cock* position of hammer of a gun about half retracted so that it cannot be operated by a pull on the trigger (Abzughahn in Ruhrast, halb gespannt); *to go off at half-cock* to go away offended and furious, but without exploding (weggehen und die Wut in sich hineinfressen).

38 26 *bureau* ['bjuərou] writing-desk

27 *wrapper* cover

35 *temper* self-control

38 *ruleth* and *taketh* show antiquated endings, Bible English

39 *Christian Scientist* a believer in *Christian Science*, a system of healing disease of mind and body, which teaches that all cause and effect is mental, and that sin, sickness, and death will be destroyed by a full understanding of the Divine Principle of Jesus's teaching and healing. The system was founded by Reverend Mary Baker Glover Eddy, of Concord, New Hampshire, in 1866, and bases its teaching on the Scriptures.

39 2 *quarterly* review or magazine published every three months

5 *shotgun* often double-barrelled, for killing small game, esp. birds (Schrotflinte)

14 *nothing much* nothing in particular

17 *squaw* Indian's wife

18 *pneumonia* [njuː'mounjə] inflammation of lungs

19 *to take it out* to pay

21 *spring* (Feder)

29 *intentionally* on purpose

31 *dresser* (Am.) chest of drawers for toilet purpose (Frisier-Kommode)

36 *porch* covered approach to entrance of building (Veranda)
screen partition; *screen door* (Windfang-, Insektenschutztür)
to slam to shut with a loud bang

37 *to catch one's breath* or 'to hold one's breath in fear' (den Atem anhalten).

40 5 *hemlock* (~ *spruce*) (Schierlingstanne)

14 *there's* there are.

OLD MAN AT THE BRIDGE

18 *pontoon bridge* temporary bridge supported by boats

22 *to grind, ground, ground* here: to move up the steep bank grinding through the dusty soil (knirschen)

27 *the enemy* here meaning 'the Fascists' (the story was written during the Spanish Civil War, described in 'For Whom the Bell Tolls')
32 *San Carlos* (*de la Rapita*) town in North-Eastern Spain, south of Tortosa.

41 10 *Ebro* river in the north-east of Spain
12 *to signal* to indicate, show
13 *contact* enemy contact (Feindberührung)
25 *to state* to mention
33 *to fork* to diverge, branch off (abzweigen)
34 *Tortosa* where the Ebro flows into the Mediterranean.

42 1 *blank* expressionless
25 *overcast* cloudy.

THE UNDEFEATED

43 11 *taxidermist* one who prepares and mounts skins of animals in lifelike manner (Tierausstopfer)
12 *poster* placard displayed in public place
17 *corrida* (Span.) bull-fight
29 *kid* (Am.) chap
31 *vara* (Span.) bull-fighter's lance
caballo (Span.) horse
32 *novillero* (Span.) young bull-fighter
34 *lot* number, quantity

44 3 *coleta* (Span.) bull-fighter's pigtail (Zopf)
to pin to fasten with pins
25 *nocturnal* night show
28 *to substitute for* to take the place of
33 *to draw* to attract lookers-on
38 *stuff* strength, energy (das ‚Zeug').

45 1 *novillo* (Span.) young bull
2 *Charlots* bull-fighters' name
4 *corral* pen, enclosure (Einfriedigung, Pferch)
5 *to pass* to let through
9 *appeal* call, request (Appell)
32 *cuadrilla* (Span.) group of bull-fighters
33 *there's* there are
nights at night, in the evening
35 *picador* (Span.) bull-fighter with lance on horseback
36 *not much* no good, not worth much.

46 2 *duro* (Span.) five pesetas
14 *even break* (coll.) chance
20 *charlotada* Spanish version of the bull-fight of the Charlots
22 *to feel* to think
30 *to pull tight* to close, shut
35 *Puerta del Sol* (Span.) Gate of the Sun
37 *to intersect* to cross one another.

47 5 *in back* at the back, behind
12 *shot* portion or certain amount of drink (Schuß)
ordinary common brandy
18 *spout* opening through which liquid is poured out
25 *you fighting?* are you fighting?
29 *you?* are you?
the Charlie Chaplins name of bull-fighters' clown-like show.

48 3 *cogida* (Span.) wound inflicted by bull
4 *wrapper* paper covering
10 *to slop* to spill (verschütten)
12 *hurt bad* instead of grammatically correct 'badly hurt'
14 *a hell of a lot he cares* (coll.) he does not care a bit
16 *to stand in with* to be a partner with, to be in secret, profitable alliance with (mit jdm. unter einer Decke stecken, gemeinsame Sache machen mit)
22 *bird* (coll.) fellow, chap
24 *ain't* isn't
31 *that bunch of camels* (coll.) all those idiots
33 *I seen* (sl.) I saw
34 *giraffe* [dʒiˈrɑːf] tall fellow (‚langes Laster').

49 18 *laborious* [ɔː] not fluent, toilsome (mühselig)
35 *manos duros* (Span.) hard hands.

50 15 *to figure* (Am.) to think
16 *to get away with* to manage
26 *I got to* I have to
33 *to go good* (sl.) to do well
36 *to go great* (sl.) to do wonderfully.

51 1 *faena* (Span.) work
4 *to get going* to get started (in Schwung kommen)
10 *to offer* to start talking
12 *reproachful* scolding, protesting (vorwurfsvoll)
16 *plenty* plenty of times
18 *to quit* to give up, leave, abandon

21 *to go big* (sl.) to do very well
22 *see?* do you understand?
24 *to relieve* to satisfy
25 *monkey business* (Am.) nonsense (Blödsinn, Unfug)
28 *stuff* here: disposition, capacity (Veranlagung, ‚Zeug')
29 *to argue* ['ɑ:gju:] to treat matter by reasoning (sich streiten).

52 1 *patio* (Span.) inner court (Innenhof)
7 *to smell good* 'to smell' is used as a copulative verb, so that it takes a predicative adjective and not an adverb
10 *you ever seen?* did you ever see?
11 *to loom* to be seen in vague and often magnified or threatening shape (verschwommen und dräuend auftauchen)
17 *arc-light* (Bogenlampe)
plaza (Span.) (market-) place, here: *plaza de toros* arena
19 *tramp* person who tramps the roads in search of work (Landstreicher, Strolch)
20 *bell-boy* (Am.) hotel page
21 *cane* walking-stick
26 *jingle* mingled noise like that of small bells or links of chain (Kettengerassel)
27 *to hitch* to fasten with loop or hook
29 *burlesque* [bə:'lesk] comedy (Posse)
barrera (Span.) barrier.

53 3 *that one* that joke
8 *following* body of adherents, followers
16 *paseo* (Span.) entering of bull-fighters into arena
19 *cape* short sleeveless coat
to furl to roll up
21 *push-pole* lance
23 *it's a wonder* it's surprising (man muß sich wundern)
26 *skin* starved old horse
31 *to sit a horse* (ein Pferd reiten)
gaunt lean, haggard, very thin
33 *to wheel* to swing round on hind legs.

54 1 *to quilt* [kwilt] to pad (polstern)
4 *matador* (Span.) man appointed to kill bull in bull-fight
5 *peon* (Span.) bull-fighter on foot
6 *in back of him* behind him
7 *Madrileno* (Span.) man from Madrid
8 *aloof* distant, away from the others (zurückhaltend, stolz)
31 *jerky* with sharp, sudden movement.

55 2 *Joselito and Belmonte* names of famous Spanish bull-fighters
4 *toril* (Span.) pen where bulls are kept (Stierzwinger)
10 *to cut down* to cause to fall; *to cut down to size* to pic the bull so that he falls and can be given the rest
16 *box-stirrups* stirrups in shape of boxes
17 *buckskin* leather made of buckskin (Wildleder)
armour protective covering (Rüstung, hier: Beinschiene)
21 *to quiver* to tremble
24 *to skid* to slide on slippery ground
25 *to charge* to attack impetuously
27 *to woof* to snort (schnauben)
31 *El Heraldo* 'The Herald'
Campagnero name of bull
32 *negro* (Span.) black
33 *gas* steam (Dampf) — The actual bull-fight is starting now. According to old rules it proceeds in four stages: (1) The bull is irritated and infuriated by red cloths (*capas*) waved by the so-called *capeadores* or *chulos;* (2) The *picadores* (Lanzenreiter) pic the bull with their lances, three times as a rule; (3) The *banderillero* plants the *banderillas*, sticks with fishhooks and decorated with ribbons, into the bull's neck; (4) The *torero* or *matador* infuriates the bull again with a scarlet cloth (*muleta*) and kills him by a shot (Stich, Stoß) with the *espada* (Stoßdegen) between the shoulder blades. The different stages of the bull-fight in this story can be found out quite easily as they are pointed out by the *El Heraldo critic*
36 *to pivot* ['pivət] to turn round sharply.

56 1 *to vault* to leap while resting on hands
6 *cash customer* paying visitor
14 *to brace* to support oneself
15 *scramble* eager struggle
23 *to billow* to move in billows or waves
34 *second-string* substitute, second-rate
36 *veronica* special trick with red cloth
Belmontistic as practised by the famous bull-fighter Belmonte
37 *recorte* (Span.) part of show (Ausschnitt)
regular regular visitor
38 *tercio* (Span.) the third; *we entered the tercio of the cavalry* we started the first third part of the bull-fight in which the fighters on horseback, the *picadores*, take the leading part. Out of the four regular parts of the bull-fight Hemingway makes three, obviously not counting the first, the waving of the capas, as a part by itself.

57 6 *to paw* to strike with hooves

19 *to gore* to pierce with horn
25 *to worry* to become afraid
28 *tomar!* (Span.) take it
35 *rosinante* (Span.) (Don Quichote's) horse (Schindmähre)
quites (Span.) attempt of diverting (Ablenkungsmanöver)
38 *to resurrect* to revive practice of (auffrischen)
39 *suerte* (Span.) round (Gang).

58 **1** *olé!* (Span.) bravo! well done!
12 *to relax* to allow to become loose or slack, to loosen (lockern)
lock grip
13 *to rip in* to make long cut or tear in (aufschlitzen)
32 *the thirds* in a bull-fight, according to Hemingway: the pic-ing by the *picadores*, the planting of the banderillas, and the killing by the *matador*; *they'll change the thirds on us* probably means that they (the managers or perhaps also the shots of the *picadores*) may change or alter the two remaining thirds of the bull-fight. Manuel thinks that by Zurito's pic-ing the bull has already been so badly hit that he may go down before the fixed time, so that the two remaining thirds of the show, the planting of the *banderillas* and the waving of the *muleta* in the last third, will either have to be shortened or cancelled altogether in order to give Manuel the chance of dealing the bull the mortal blow
34 *I got to go* I have to go
35 *mono* (Span.), *mono sabio* the picador's assistant
36 *to whack* to strike heavily with stick.

59 **2** *to scowl* to look sullen (finster blicken)
6 *Zurito watched* ... Notice the hacked-off sentences, illustrative of the excitement of the scene
14 *to lung* to rush, make sudden start
17 *sheen* brightness (Schein, Glanz)
30 *awful close* awfully close (see note to page 10)
that close so close
39 *toro* (Span.) bull.

60 **3** *swirl* whirling motion (Wirbel)
4 *pass* here: movement
10 *banderillo* (Span.) mixed up by Hemingway with *banderilla* stick with fish-hooks (mit Widerhaken versehener Spieß)
24 *champagne* [ʃæm'pein] (Sekt)
26 *to rate* to receive
27 *the third of the palings* the third of the bull-fight in which the *banderillas* are planted in the bull's neck by *banderilleros*; *paling* really means a

fence of pales, of pointed pieces of wood. The newspaper critic wants to say by *palings* that the fish-hook sticks, the *banderillas*, planted in the bull's neck one beside the other, look like a *paling*, a fence of pointed sticks

37 *to twitch* to pull with light jerk.

61 12 *trout* freshwater fish (Forelle)

14 *to toss* to be thrown up

30 *stuff* affair, thing, matter

32 *apprehension* worry

34 *to reduce* here: to weaken

35 *manageable* so that you can finish him

37 *to walk heel-and-toe* to move feet from heel to toe as in walking-races.

62 9 *the crowd were wild* the plural *were* indicates individuality of the crowd, the individual way in which each spectator reacted

10 *stuff* show

18 *barrel* belly

to heave to swell, rise up and down

25 *barb* recurved point of arrow or fish-hook (Widerhaken); *barbed wire* (Stacheldraht).

63 7 *to tuck* to wrap, cover up (in)

baton ['bætən] staff, stick

8 *muleta* (Span.) scarlet cloth

10 *scabbard* sheath of sword (Scheide)

12 *hilt* handle

limp not stiff, flexible (schlaff)

17 *just like* (Am.) just as

27 *to get* to hear, learn

28 *date* appointment (Verabredung)

29 *Maxim's* name of fashionable restaurant

30 *bum* (Am.) loafer, vagabond.

64 14 *to go in* to push the sword in

18 *successively* one after another

26 *jib* triangular sail (Klüver, dreieckiges Vorsegel)

27 *to splinter* to split into long thin pieces (zersplittern)

28 *porcupine* ['pɔːkjupain] animal covered with spikes all over (Igel)

29 *quill* [kwil] spine, spike

33 *to reserve oneself* to rest and gather strength

34 *I've got to* I must.

65 5 *to tighten* to narrow

11 *pass* rushing past

23 *pase natural* (Span.) natural posture of bull-fighter
24 *pase de pecho* (Span.) posture of chest (Bruststellung)
25 *to come by* to pass.

66 1 *to dig up* to discover, find
4 *damn quick* (vulgar) very quickly
7 *to kid* (Am.) to make fun
23 *to bump oneself up* to rise clumsily and with great effort
31 *to motion* to give signs.

67 7 *lead* [e] a metal (Blei)
17 *gone* lost
20 *to profile oneself* to move quickly
25 *pitch* incline, climb, rising (Steigung, Buckel)
28 *corto y derecho* (Span.) short and straight
34 *to sight* to look
to dip to lower (sich senken)
36 *to launch* to throw.

68 2 *slippered* with slippers (mit dünnen Schuhen — eigentlich Pantoffeln — bekleidet)
39 *the small of his back* the hinder part of his waist (Kreuz).

69 4 *to try* to test (prüfen)
21 *to go in* to attack.

70 2 *to chop* to change direction suddenly
3 *nothing doing* (coll.) it was no use
5 *to buckle* to curve, bend
7 *end-over-ending* turning round all the time
11 *to spot* to cover here and there
13 *from close range* from very near
15 *to whisk* to fly
by him near him
27 *to jam* to squeeze, thrust violently.

71 1 *infirmary* hospital
11 *guard* contrivance to prevent injury (Schutzvorrichtung)
14 *to lurch* to stagger, stumble
37 *to scald* [ɔː] to burn, boil (verbrühen).

72 27 *to grab* to seize.

73 6 *cone* here: kind of cap in shape of a cone (Kappe)
7 *awkward* clumsy, embarrassed (unbeholfen).

D E F G H ——————— V W X Y Z